Optimizing Predictive Maintenance of Biomedical Equipment

I dedicate this work to the soul of my dear friend, Rabii Ben Kadissa, hoping that God will have mercy on him and grant him paradise.

Je dédie cet ouvrage à l'âme de mon cher ami Rabii ben Kadissa, en espérant que Dieu aura pitié de lui et le fera habiter au Paradis.

أهدي هذا العمل إلى روح صديقي العزيز ربيع بن كديسة راجيا من الله أن يرحمه و يسكنه الجنة

Reliability of Multiphysical Systems Set

coordinated by
Abdelkhalak El Hami

Volume 22

Optimizing Predictive Maintenance of Biomedical Equipment

Faker Bouchoucha

iSTE

WILEY

First published 2026 in Great Britain and the United States by ISTE Ltd and John Wiley & Sons, Inc.

ISTE Ltd
27-37 St George's Road
London SW19 4EU
UK

www.iste.co.uk

John Wiley & Sons, Inc.
111 River Street
Hoboken, NJ 07030
USA

www.wiley.com

Library of Congress Control Number: 2026933534

British Library Cataloguing-in-Publication Data
A CIP record for this book is available from the British Library
ISBN 978-1-83669-120-4

The manufacturer's authorized representative according to the EU General Product Safety Regulation is Wiley-VCH GmbH, Boschstr. 12, 69469 Weinheim, Germany, e-mail: Product_Safety@wiley.com.

Contents

Preface

Building a Bridge between Engineering and Healthcare through Predictive Maintenance

In an era where healthcare systems face growing demands for efficiency, safety and sustainability, the role of biomedical engineering has never been more critical. This book is born out of a deep conviction that science and engineering can – and must – contribute directly to improving patient care. It reflects years of research, academic engagement and practical observation within healthcare environments, where the reliability and availability of medical equipment determine life-or-death outcomes.

Focusing specifically on dialysis machines – among the most demanding and vital pieces of medical equipment – this work seeks to demonstrate how advanced maintenance strategies rooted in data analytics, reliability theory and probabilistic modeling can make a tangible difference. By integrating tools such as Weibull analysis, fault tree analysis and Bayesian networks, we aim to move beyond reactive and preventive maintenance toward a predictive paradigm that anticipates failure before it occurs.

This book is designed for a broad audience – engineers, healthcare professionals, researchers and decision-makers – who share a common goal: to ensure that medical technologies function reliably, safely and sustainably in service of patient care. The chapters are structured to guide the reader step-by-step, from theoretical foundations to practical applications, always with the intention of bridging the gap between technical modeling and real-world implementation.

Beyond the academic and technical dimensions, this book is also a tribute – to colleagues, mentors and especially to the memory of a dear friend whose support and humanity continue to inspire. It is my sincere hope that the knowledge shared in these pages will not only contribute to the advancement of biomedical maintenance strategies but also serve as a reminder that behind every machine is a patient, and behind every maintenance strategy is a human life we strive to protect.

March 2026

Abstract

In today's healthcare landscape, biomedical equipment plays an essential role in delivering high-quality patient care, particularly in environments where reliability, safety and service continuity are paramount. Dialysis machines are among the most maintenance-intensive and life-critical devices, which require consistent performance to ensure patient survival and treatment effectiveness. However, hospitals often face significant challenges in maintaining this equipment due to limited budgets, the complexity of devices and the inadequacy of traditional maintenance strategies. Current maintenance approaches – largely corrective or preventive – fail to address the growing need for predictive, data-driven methods that can anticipate failures and optimize resource allocation.

To meet this demand, healthcare institutions are increasingly turning to reliability-centered and data-driven maintenance strategies. These strategies focus on improving two core performance indicators: reliability, defined as the likelihood of equipment functioning without failure over a specified period, and availability, the proportion of time equipment is ready for clinical use. The effective management of these metrics directly influences patient outcomes and institutional efficiency. In many healthcare facilities, particularly in developing regions, the absence of structured maintenance policies contributes to high rates of equipment failure, as evidenced by World Health Organization (WHO) reports indicating that over 50% of medical devices may be non-functional.

Central to predictive maintenance is the systematic use of failure history and performance data. Key metrics such as mean time between failures (MTBF), mean time to repair (MTTR) and utilization rates can guide more informed and proactive maintenance schedules. In dialysis units, where failure of components like water treatment systems or control modules can have critical consequences, analyzing historical maintenance logs enables hospitals to target their interventions effectively and reduce equipment downtime.

To prioritize these interventions, Pareto analysis is employed to identify the most frequent and impactful failure types based on the principle that 80% of problems stem from 20% of causes. This method allows maintenance teams to allocate resources efficiently by focusing on high-priority components. Complementing this, fault tree analysis (FTA) provides a deductive approach to understanding the causal relationships between system components and failures. By modeling how different failure modes contribute to a critical event, such as a dialysis machine malfunction, FTA facilitates both qualitative insight and quantitative risk assessment.

While FTA is suitable for well-defined systems, it has limitations in dynamic and uncertain environments. To overcome this, Bayesian networks (BNs) are introduced as powerful probabilistic models that can incorporate diverse data sources, including sensor outputs, expert knowledge and historical records. BNs enable real-time prediction of equipment failure and allow for continuous updates as new data becomes available. For example, by integrating real-time flow, temperature and pressure data from a dialysis machine, BNs can assess the probability of a pump failure and recommend timely interventions.

Addressing uncertainty is crucial in maintenance planning. Parameters such as environmental stressors, usage patterns and human error introduce variability that deterministic models cannot capture. Stochastic modeling, including Monte Carlo simulations, provides a framework for evaluating a wide range of potential outcomes and reinforcing the robustness of maintenance policies. Additionally, statistical tools like the Weibull distribution are instrumental in modeling the life expectancy and degradation behavior of medical devices, supporting optimized maintenance intervals and lifecycle management.

Implementing an integrated, data-driven maintenance framework requires collaboration among biomedical engineers, IT personnel and clinical stakeholders. Core components include centralized failure databases, real-time monitoring via Internet of Medical Things (IoMT), predictive analytics engines and risk assessment tools. When supported by a robust Computerized Maintenance Management System (CMMS), such frameworks enhance decision-making, improve equipment reliability and ultimately contribute to safer and more cost-effective healthcare delivery.

In conclusion, transitioning from reactive to predictive maintenance through data analytics, probabilistic modeling and reliability-centered methods represents a transformative opportunity for healthcare institutions. This shift not only enhances the performance of biomedical equipment like dialysis machines but also aligns technical strategy with ethical imperatives – ensuring that equipment failures do not compromise patient care.

Organization of the Book

This book is structured into seven interrelated chapters, each building progressively toward the development of a robust predictive maintenance strategy for dialysis machines based on reliability analysis and probabilistic modeling.

– *Chapter 1: Medical Equipment Management in Healthcare Facilities*

This introductory chapter presents an overview of medical equipment management practices within healthcare environments. It discusses the organizational, regulatory and technical challenges related to the acquisition, operation and maintenance of biomedical devices. The focus is placed on the need for structured maintenance planning to ensure equipment safety, availability and cost-effectiveness.

– *Chapter 2: Description of the Dialysis Machine*

This chapter provides a comprehensive technical description of the dialysis machine. It outlines its main components, working principles and critical subsystems such as the hydraulic circuit, filtration modules and control interfaces. The aim is to identify the machine's vulnerability points and maintenance-critical components, which will be analyzed in subsequent chapters.

– *Chapter 3: Weibull Distribution*

This chapter introduces the Weibull distribution as a versatile statistical model for analyzing time-to-failure data. The theoretical background, parameter estimation techniques and interpretation of shape and scale parameters are presented. The relevance of this distribution for modeling the failure behavior of medical devices is emphasized.

– *Chapter 4: Reliability, Maintainability and Availability Analysis through Weibull Distribution*

In this chapter, the Weibull model is applied to analyze the reliability, maintainability and availability (RMA) of generic medical equipment. This chapter develops the mathematical foundations of RMA metrics and demonstrates how they can be derived from real or simulated failure data. Graphical representations and practical examples are used to illustrate the relationships between system behavior and its Weibull parameters.

– *Chapter 5: Reliability, Maintainability and Availability of the Dialysis Machines based on the Failure History Through Weibull Distribution*

This chapter focuses the RMA analysis on dialysis machines. Using real-world or representative failure datasets, it identifies the most failure-prone components, estimates their Weibull parameters and interprets the implications for maintenance planning. The results allow for prioritizing of maintenance actions and optimizing service intervals to improve machine availability and reduce downtime.

– *Chapter 6: Effect of the Stochastic Weibull Parameters on the Reliability of the Dialysis Machine*

In this chapter, a methodology is introduced to evaluate how influencing factors affect system reliability. The approach involves applying controlled variations to the shape and scale parameters of the Weibull distribution to analyze their impact on key reliability metrics. The Weibull parameters are estimated analytically using historical failure data from the equipment. These parameters are treated as stochastic variables assumed to follow a Gaussian distribution. The mean and standard deviation for both the shape and scale parameters are computed.

Subsequently, computational models are used to generate statistical insights into the probabilistic reliability behavior of the dialysis machine.

– *Chapter 7: Fault Tree and Bayesian Network-Based Probabilistic Modeling for Predictive Maintenance of Dialysis Machines*

This chapter presents a dual probabilistic modeling approach combining fault tree analysis (FTA) and Bayesian networks (BNs). FTA is used to identify and visualize failure logic pathways, while BNs offer a dynamic, data-driven model for updating failure probabilities in real time. The integration of both tools forms the basis for an intelligent predictive maintenance framework that enhances decision-making under uncertainty.

Each chapter contributes to the overall objective of this work: to develop a scientifically grounded and practically applicable predictive maintenance model for dialysis machines, ensuring higher reliability, safety and operational efficiency in healthcare settings.

Introduction

Hospitals and healthcare institutions represent some of the most complex and technologically demanding environments in modern society. They bring together diverse medical disciplines, specialized staff and advanced technologies, all of which must operate under the critical obligation of delivering safe, efficient and high-quality care. In such a context, biomedical equipment plays a central role. From basic diagnostic tools to sophisticated life-support devices such as dialysis machines, the reliability, performance and availability of equipment are not simply operational considerations but fundamental determinants of patient safety and institutional effectiveness. Ensuring that these technologies function as intended is therefore both a technical and ethical responsibility.

Reliability in biomedical engineering refers to the probability that a device performs its required function without failure for a specified period, while availability measures the proportion of time it remains operational and ready when needed. In healthcare settings, the two are inseparable: a device that frequently fails cannot be considered available, and reduced availability inevitably compromises care delivery. The importance of this relationship is underscored by global challenges. According to historical data from the World Health Organization (WHO), over 50% of medical equipment in developing nations was found to be non-operational as a result of insufficient maintenance practices, a challenge that continues to be a significant concern in the present day (World Health Organization, Regional Committee for the Eastern Mediterranean 2006). This makes clear that maintenance is not optional but indispensable for ensuring safe and effective healthcare services. Dialysis machines vividly illustrate this reality. Since patients depend on them several times a week for life-sustaining treatment, even minor interruptions in availability can have severe clinical consequences. They thus serve as a paradigm for examining the broader challenges of biomedical maintenance.

Healthcare institutions face significant difficulties in organizing effective maintenance strategies. The large variety of devices complicates standardization, while limited budgets often reduce investment in preventive measures. Furthermore, traditional strategies tend to be either corrective, reacting after failures occur, or preventive, based on fixed time intervals without regard to actual usage. These approaches are costly and often inefficient in the high-stakes environment of modern hospitals. Predictive maintenance, which is based on numerical data to anticipate failures before they occur, offers greater potential but remains underused due to the complexity of modeling degradation and managing uncertain factors such as user behavior and environmental conditions. Moving toward data-driven and probabilistic models offers a way to address these challenges.

One of the most valuable yet underexploited resources in this context is historical failure data. Maintenance logs, performance records and incident reports contain crucial information about degradation patterns and recurrent problems. A systematic examination of these data enables the calculation of essential performance metrics, including mean time between failures (MTBF) and mean time to repair (MTTR), which in turn inform strategic maintenance planning, optimize resource distribution and shape targeted training programs for personnel. In dialysis units, for example, recurring issues with pumps, water systems or control circuits can be identified and addressed more effectively when informed by past records. This transforms maintenance from a reactive process into a proactive and knowledge-driven strategy.

As hospitals operate with limited resources, prioritizing interventions is essential. Pareto analysis provides a practical way to focus on the most critical problems. Following the principle that 80% of failures often arise from 20% of causes, this method identifies the equipment or components that generate the most downtime. Applied to dialysis units, for instance, it may reveal that temperature control malfunctions or saline infusion pump issues account for most service disruptions. By targeting these high-impact areas, institutions can significantly improve availability with relatively limited effort. Pareto charts make these insights visually accessible, helping managers allocate attention and resources effectively.

Beyond prioritization, understanding how failures occur requires structured analysis. Fault tree analysis (FTA) is particularly useful, offering a deductive approach that begins with an undesirable event, such as the sudden failure of a dialysis machine, and works backward to identify possible causes. Through logical relationships expressed with AND and OR gates, FTA reveals how different components interact and where single points of vulnerability exist. Assigning probabilities to these causes enables quantitative risk assessments, guiding decisions about redundancy, inspections and design improvements. This structured method is especially valuable in regulated environments such as dialysis care, where patient safety cannot be compromised.

Nevertheless, static models like fault trees have limitations in dynamic or uncertain contexts. Bayesian networks provide a more flexible probabilistic framework, allowing for the integration of historical data, expert knowledge and real-time sensor inputs. They not only estimate the likelihood of failures under varying conditions but also update predictions as new data become available. A dialysis machine equipped with flow, pressure and temperature sensors, for example, can be monitored through a Bayesian model that continuously refines its estimates of pump failure risk. This capacity supports predictive maintenance in real time, enabling timely interventions that reduce downtime and protect patients. Moreover, Bayesian models can also reflect the impact of completed maintenance actions by recalculating reliability after repairs or adjustments, creating a continuously optimized cycle.

Uncertainty is a central challenge in all of these approaches. Device use varies among patients and departments, environmental conditions change and human error is inevitable. Deterministic models cannot adequately capture this variability. Stochastic methods are therefore needed, capable of accounting for random measurement errors and poorly known parameters like degradation rates, variable environmental factors and the inherent simplifications of system models. Monte Carlo simulations are widely used in this regard, generating thousands of possible scenarios to estimate failure probabilities across a range of conditions. This ensures that maintenance strategies are robust, adaptable and effective, even under unpredictable circumstances.

Statistical tools also contribute significantly to biomedical maintenance. Among them, the Weibull distribution is extensively employed across numerous fields for modeling failure behavior across different stages of a device's lifecycle. Its flexibility allows it to represent increasing, constant or decreasing failure rates, making it especially relevant for complex systems like dialysis machines. By applying Weibull analysis, hospitals can better interpret historical data, forecast failures and determine optimal maintenance intervals. When combined with other methods such as fault tree or Bayesian models, Weibull analysis strengthens predictive accuracy and helps build proactive maintenance schedules that extend device lifespan and improve patient safety.

Bringing these tools together requires an integrated, data-driven framework. At the heart of such a framework lies a computerized maintenance management system (CMMS), capable of collecting, processing and visualizing data. A comprehensive system would include a centralized database for maintenance records, real-time monitoring through Internet of Medical Things (IoMT) sensors, predictive modules based on Bayesian networks and risk assessment tools built on FTA. Optimization algorithms could further balance competing objectives, such as cost, reliability and availability. Implementing such a system demands collaboration among biomedical engineers, IT specialists, clinicians and administrators, as well as continuous training and auditing to ensure it remains effective as technology and clinical practices evolve.

Ultimately, as healthcare grows increasingly dependent on complex technologies, ensuring the reliability and availability of biomedical equipment is more critical than ever. Dialysis machines exemplify the stakes involved: they are life-sustaining devices that highlight both the challenges of maintenance and the transformative potential of advanced strategies. By moving from reactive to predictive and reliability-centered maintenance through the integration of failure history, Pareto prioritization, fault tree and Bayesian modeling, and stochastic simulations, hospitals can improve equipment uptime, reduce costs and, most importantly, enhance patient safety. This transition represents not just a technical upgrade but a cultural and organizational shift, one in which data are transformed into insight, and insight into better decisions. In this way, healthcare institutions can align their maintenance practices with their ultimate mission: delivering safe, reliable and high-quality care that supports and protects the lives of their patients.

1

Medical Equipment Management in Healthcare Facilities

1.1. Introduction

Healthcare facilities, whether public hospitals, private clinics, or specialized centers, are at the heart of the modern care system. Their core mission is to ensure access to quality, safe and equitable care in environments where human, material and informational resources must coherently interact. In this context, medical equipment, which forms the technological backbone of care delivery, plays a strategic role. It enables healthcare professionals to diagnose accurately, treat effectively and continuously monitor patients (David and Jahnke 2004).

Healthcare facilities, from major public hospitals to private clinics, constitute the cornerstone of the modern healthcare system. Their core mission is to ensure access to high-quality, safe and equitable care within an environment where human, material and informational resources must function in a coherent and integrated manner. Medical equipment, serving as the technological backbone of care delivery, plays an essential strategic role. It supports accurate diagnosis, effective treatment and continuous patient monitoring. Moreover, the growing connectivity of these devices – often referred to as the Internet of Medical Things (IoMT) – positions them as vital nodes within the broader healthcare information ecosystem.

This increasing reliance on technology, however, introduces critical vulnerabilities. Equipment failure is no longer merely a technical inconvenience; it represents a potential rupture in the care continuum, leading to diagnostic delays, therapeutic interruptions and compromised patient safety. Traditional maintenance models, which tend to be reactive or based on fixed schedules, have proven to be both inefficient and costly.

To address these challenges, our research proposes a sophisticated predictive maintenance methodology aimed at transforming biomedical asset management. This approach is built upon three foundational pillars. First, Weibull analysis leverages historical failure data to transition from calendar-based to condition-based maintenance (CBM) using statistical modeling to estimate failure probabilities and remaining useful life. Second, fault tree analysis (FTA) provides a deductive, hierarchical framework for identifying all potential technical, human and procedural causes of failure, enabling a deeper understanding of root causes. Finally, Bayesian networks (BN) integrate these elements into a dynamic intelligence system, continuously updating failure risks in real time, based on incoming data from sensors and performance tests.

The operational goal of this integrated methodology is to enable proactive and precisely targeted maintenance. In practice, the system will generate early warnings – such as "component Y has a 92% probability of failure within the next fifteen days" – allowing interventions to be scheduled during periods of low clinical activity, thereby minimizing disruption.

The benefits extend across the entire healthcare ecosystem. Facilities will achieve greater equipment availability, optimized maintenance budgets and more efficient allocation of technical resources. Clinicians will gain confidence and operational serenity and be able to focus on patient care without unexpected technical interruptions. Most importantly, patients – the ultimate beneficiaries – will experience enhanced safety and care continuity, contributing to better health outcomes and a more reliable care experience.

In summary, this modernization of maintenance practices transcends technical improvement – it supports the broader objective of strengthening the resilience and performance of the entire healthcare system. By making technology a dependable and efficient lever, we put it firmly at the service of human health.

1.2. The central role of medical equipment in care delivery

Medical devices are a fundamental pillar of modern medicine, covering an extremely broad spectrum of tools and technologies. This range extends from surgical instruments and radiology equipment to laboratory devices, physiological monitors and infusion pumps. Life-support equipment, such as defibrillators, incubators and ventilators, represents a crucial category for maintaining vital functions. Today, this physical dimension is complemented by a growing digital dimension, with the integration of solutions such as electronic health records and telemedicine, which are revolutionizing health data management (World Health Organization 2025). The presence of these technologies is now ubiquitous and systemic, spanning the entire patient care pathway. Their use begins in emergency departments for diagnosis and immediate stabilization, continues in intensive care units for continuous monitoring and life support, and extends into rehabilitation departments, where they play a key role in recovery and long-term follow-up.

Technological advances have profoundly transformed the hospital ecosystem, notably bringing about a revolution in task automation, diagnostic accuracy and rapid intervention. However, this growing sophistication, while synonymous with progress, introduces operational complexity that requires rigorous management and dedicated expertise. Indeed, medical devices, although designed to save lives, can paradoxically become the source of critical clinical failures. The slightest technical failure, a human–machine interface error, or a targeted cyberattack can compromise patient safety, delay urgent care or, in the worst case, lead to medical errors with dramatic consequences (Arya et al. 2019).

Beyond purely technical risks, this increased dependence on technology raises major human and organizational challenges. It requires ongoing training for caregivers to ensure perfect mastery of the tools, demanding preventive maintenance, and seamless interoperability between systems to guarantee continuity and traceability of care. Thus, the modern hospital must find a delicate balance: fully leveraging the innovative potential of technology while mitigating the risks it generates, so that progress always translates into enhanced patient safety.

The proper functioning of these devices depends on multiple factors: maintenance, component quality, compatibility with hospital software and regular firmware updates. Moreover, increased reliance on digital equipment introduces new vulnerabilities, particularly cybersecurity risks, which must be integrated into the hospital's overall risk management strategy (Tully et al. 2020).

The reliable and secure operation of these devices depends on a multitude of interdependent factors, forming a complex technical and organizational ecosystem. Critical elements include stringent preventive and corrective maintenance protocols, high intrinsic quality and full traceability of components, seamless integration and compatibility within hospital software architectures, and the consistent and timely application of software and firmware updates – essential for addressing vulnerabilities and sustaining performance.

Moreover, growing dependence on interconnected digital equipment introduces emerging risks, notably in the domain of cybersecurity. Threats such as ransomware, data breaches and attacks on critical systems cannot be addressed in isolation. Proactive and continuous management of these risks must be comprehensively and cross-functionally embedded into the healthcare facility's overall risk management strategy. Such integration is vital in ensuring the resilience of the hospital information system, safeguarding patient data confidentiality and ultimately upholding the continuity and safety of care in an increasingly digital environment (Tully et al. 2020).

1.3. Organization and responsibilities of the biomedical maintenance department

The integrity of medical equipment, which is fundamental to patient safety and the uninterrupted delivery of care, is entrusted to a highly specialized unit known as the Biomedical Engineering Department. More than just a technical support team, this department functions as a multidisciplinary hub where biomedical engineers and technicians integrate expertise across electronics, information technology, precision engineering and healthcare regulations. Its responsibilities extend well beyond remedial repairs to include holistic management of the medical device inventory. This encompasses technical activities – such as performing device calibrations, software upgrades and hardware interventions – as well as organizational functions like systematic scheduling of preventive maintenance and maximizing the operational readiness of critical care equipment. In addition, the department maintains comprehensive documentation, including service records, compliance certificates and full traceability of components. This ensures not only reliable equipment performance but also adherence to stringent regulatory standards and the capacity to review all actions for future audits or incident analyses. Through its integrated approach to technical, operational and documentary oversight, the biomedical engineering department serves as an indispensable component in the resilience and efficiency of the healthcare system.

Medical equipment maintenance is handled by a specialized unit: the biomedical maintenance department. This department manages the technical, organizational and documentation aspects of the biomedical inventory. Its key responsibilities include the following:

– *Inventory management*

The implementation of an intelligent, centralized and relational database enables comprehensive traceability and advanced strategic analysis of medical equipment. This sophisticated system integrates key parameters such as device category and classification, precise physical location (department, unit, floor, building), detailed manufacturer information including technical support contacts and service level agreements, acquisition dates, commissioning timelines, warranty status with expiration tracking and a complete historical record of all interventions – including preventive maintenance, corrective actions, calibration certificates, software updates and user-reported observations. This structured framework not only streamlines maintenance scheduling and optimizes equipment availability but also provides actionable, data-driven insights to support strategic decisions related to lifecycle management, fleet standardization, resource allocation and procurement planning. Advanced analytics capabilities further enhance operational efficiency by identifying usage patterns, predicting maintenance needs, and minimizing downtime through proactive intervention.

– *Preventive maintenance*

Planned in strict accordance with manufacturers' specifications and current technical standards, preventive maintenance serves as a foundational element of effective biomedical equipment management. It extends beyond routine scheduled tasks by incorporating data-driven insights from failure history analytics, usage patterns and operational context. Through systematic performance verification, comprehensive cleaning protocols, precision calibration and proactive replacement of high-risk components, this approach substantially decreases unexpected equipment failures and enhances device longevity. By maintaining devices in optimal operating condition, it improves clinical workflow efficiency, ensures consistent performance accuracy and, most importantly, safeguards patient safety by reducing the potential for equipment-related adverse events. Furthermore, it embodies a philosophy of continuous improvement: each maintenance activity contributes to an expanding knowledge repository, enabling refinement of maintenance intervals, procedures and overall asset management strategy.

– *Corrective maintenance*

Initiated in response to equipment failure or user reports, corrective maintenance responses are prioritized based on the clinical criticality and operational impact of each incident. For failures that compromise patient safety or disrupt urgent clinical procedures, immediate intervention is mandatory. For non-critical equipment or minor malfunctions, deferred maintenance may be scheduled to optimize resource allocation without affecting patient care. Successful execution depends on responsive logistical support – including strategic inventory management of spare parts, access to specialized diagnostic equipment and the availability of qualified technical staff – coupled with comprehensive documentation that cover fault diagnosis, repair actions, components replaced and equipment downtime. This systematic approach not only restores equipment function but also supports continuous improvement through root cause analysis and ensures adherence to regulatory standards and accreditation requirements (Leming et al. 2023).

– *Quality control*

Beyond routine maintenance, critical medical devices demand a rigorous regime of periodic performance verification, compliance audits and precise calibration or recalibration to guarantee both metrological integrity and clinical validity. These systematic quality assurance processes – mandated by strict international standards (such as ISO 9001, ISO 13485 and IEC 62353) and regulatory guidelines – ensure that all measured parameters (e.g. pressure, flow, radiation dosage, electrical current and temperature) remain accurate, traceable and consistent across the entire device lifecycle. Each operation is documented in detail, creating an auditable trail that captures measurement results, adjustments made and compliance status. This end-to-end traceability not only safeguards patient safety by preventing diagnostic errors and therapeutic inaccuracies but also supports institutional accreditation, reinforces legal compliance and fosters a culture of excellence in clinical engineering.

– *User training and competency assurance*

Inadequate operator technique remains a significant source of premature device degradation, preventable operational faults and adverse clinical events. To mitigate these risks, structured training programs – delivered through competency-based learning pathways including hands-on sessions, simulation-based exercises and digital learning platforms – are systematically developed and integrated into the clinical technology management lifecycle. These educational interventions address essential operational protocols (power-up, routine operation and safe shutdown), as well as advanced topics such as cleaning and disinfection procedures, early error recognition, anomaly reporting workflows and emergency response actions. Sustained and role-specific training implementation directly enhances device reliability, reduces use-related hazards, extends asset service life and strengthens clinical staff proficiency and autonomy (Kremer et al. 2023).

Implementation of an integrated Computerized Maintenance Management System (CMMS) serves as a foundational enabler for data-informed maintenance orchestration and end-to-end process traceability. These enterprise platforms provide centralized oversight and automation of the entire medical equipment support ecosystem, including real-time asset status monitoring, intelligent scheduling of preventive maintenance and calibration activities, spare parts inventory management and root-cause analysis of failure trends.

Furthermore, advanced CMMS solutions generate key performance indicators – for example, mean time between failures (MTBF), overall equipment effectiveness (OEE), maintenance cost-per-use and compliance rate – enabling evidence-based decision-making, predictive maintenance modeling, strategic resource planning and continuous enhancement of clinical engineering services in alignment with patient safety and organizational performance goals.

1.4. Regulatory and normative framework of medical devices

The medical device sector encompasses a rapidly evolving technological landscape, extending from fundamental surgical instruments to advanced artificial intelligence (AI)-based diagnostic software and connected health solutions. This dynamic field is governed by a sophisticated regulatory framework designed to balance two critical objectives: ensuring patient safety and product efficacy while simultaneously promoting innovation and timely market access. These regulations form a multi-tiered structure operating at international, regional and national levels, creating both challenges and opportunities for manufacturers and healthcare providers.

The regulatory ecosystem rests upon two fundamental pillars:

– *Regulations*

Regulations such as the European Union's Medical Device Regulation (MDR) 2017/745 and the U.S. Food and Drug Administration's (FDA) Quality System Regulation (21 CFR Part 820) establish mandatory legal requirements that devices must satisfy before and after market entry. These binding instruments govern critical aspects including product classification, conformity assessment procedures, unique device identification (UDI) systems, clinical evidence requirements, post-market surveillance and incident reporting protocols. They carry the force of law and are enforced by regulatory authorities through audits, inspections and market surveillance activities.

– *Technical standards*

Technical standards, including ISO 13485 (Quality Management Systems), IEC 60601 (Safety and Essential Performance) and ISO 14971 (Risk Management) – while typically voluntary in application – provide the necessary technical detail and methodologies to demonstrate compliance with regulatory requirements. These consensus-based documents, developed by international standardization bodies, offer precise engineering specifications, validation methodologies and best practices covering design, manufacturing, testing, risk management and documentation. Their application provides manufacturers with predictable pathways to conformity and facilitates global market access through harmonized technical requirements.

Together, regulations and standards create a coherent framework that ensures medical devices throughout their entire lifecycle, from initial concept and design through production, clinical use and eventual decommissioning, and meet consistently high standards of safety, performance and quality. This document provides a comprehensive overview of the international and European regulatory landscape, detailing the essential technical standards that govern medical devices and examining the ongoing challenges in global regulatory harmonization. It further explores how emerging technologies, such as AI-driven diagnostics, wearable medical devices and digital health applications, are reshaping regulatory paradigms and creating new requirements for evidence generation, cybersecurity and lifecycle management.

1.4.1. *Typology of regulations applicable to medical devices*

1.4.1.1. *Regulatory frameworks by geographic area*

Medical devices are subject to specific legal requirements, depending on the markets where they are commercialized. These regulatory frameworks establish the conditions for access to each market and define the legal obligations of manufacturers.

1) *Supranational regulations*

Regulation (EU) 2017/745 on medical devices (MDR) constitutes the binding legal framework for all member states of the European Union. It introduces a reinforced approach to clinical evaluation, imposes an extensive product traceability system through UDI and establishes strict post-market surveillance requirements. Its application is mandatory for any marketing within the European Economic Area.

2) *National regulations*

Each country has its own regulatory system. In the United States, the Food and Drug Administration (FDA) enforces Title 21 CFR Part 820 (Quality System Regulation), which governs manufacturing practices, as well as distinct approval processes such as Premarket Notification 510(k) for substantially equivalent devices or Premarket Approval (PMA) for innovative technologies. Japan follows the Pharmaceutical and Medical

Device Act under the supervision of the PMDA, while China enforces NMPA regulations with specific requirements for local clinical trials.

1.4.1.2. *Cross-cutting technical and process requirements*

Beyond geographical boundaries, certain technical requirements apply universally and concern all manufacturers, regardless of their location.

1) *Quality and risk management requirements*

The implementation of a system of quality management compliant with ISO 13485 has become a nearly mandatory requirement for operating in international markets. Although technically a standard, its adoption is required by regulatory authorities as proof of compliance with quality and safety principles. Similarly, the application of ISO 14971 for risk management is systematically required to demonstrate a structured approach to risk assessment and control.

2) *Traceability and identification requirements*

The UDI (Unique Device Identifier) system was made mandatory by the European MDR and US FDA regulations. It requires manufacturers to assign unique identifiers to each device, record this information in centralized databases and maintain complete traceability from production to the patient. This system not only enables targeted and effective recalls but also improves monitoring of real-world device performance.

3) *Vigilance and post-market surveillance requirements*

Modern regulations place particular emphasis on continuous post-market surveillance. The MDR requires the systematic transmission of incident reports via the Eudamed system and the regular production of periodic safety reports. The FDA maintains the MAUDE system to collect adverse event reports, while other markets such as Canada and Australia have implemented similar systems. These mechanisms enable rapid problem detection and ongoing benefit–risk assessment.

4) *Specialized requirements and recommendations*

In response to emerging technologies, authorities develop specific guidance documents. The FDA's guidance on cybersecurity of connected devices or the evaluation of AI algorithms establishes detailed technical expectations. Although not legally binding, these documents reflect the regulators' stance and have become essential references for obtaining marketing authorizations.

1.4.2. *Typology of standards applicable to medical devices*

1.4.2.1. *Quality management: the foundation*

ISO 13485:2016 serves as the cornerstone of Quality Management Systems (QMS), particularly the medical device industry. Unlike ISO 9001, which emphasizes customer satisfaction, ISO 13485 prioritizes device safety, regulatory compliance and risk mitigation. It establishes a rigorous framework to ensure that all processes, including design, installation, production, development and servicing, are consistently controlled and documented. Key applications span technical documentation, supplier qualification, production oversight and management of non-conforming products. Compliance is often a legal prerequisite for obtaining CE marking (under EU MDR) or FDA approval (under 21 CFR Part 820) and is rigorously audited by notified bodies and regulatory authorities.

1.4.2.2. *Risk management: the cornerstone of safety*

ISO 14971:2019 defines a comprehensive framework for risk management throughout a medical device's lifecycle. It guides manufacturers in systematically identifying, evaluating and controlling risks associated with device use. The process encompasses hazard analysis, risk assessment, implementation of mitigation measures, evaluation of residual risks and continuous post-market surveillance. This standard is indispensable for demonstrating to regulators (e.g. under EU MDR or FDA requirements) that a device's benefits outweigh its risks. It seamlessly integrates with ISO 13485, embedding risk-based decision-making into the QMS.

1.4.2.3. *Safety of medical electrical equipment*

The IEC 60601-1 series specifies essential requirements for the electrical, mechanical and thermal safety of medical electrical equipment. Its primary goal is to protect patients and operators from hazards such as electric shock, excessive temperatures, radiation exposure and mechanical failures. Applicable to all electrically powered medical devices (e.g. ventilators, defibrillators and patient monitors), it mandates stringent testing for insulation integrity, electromagnetic compatibility (EMC), mechanical stability and environmental resilience. Compliance is legally mandatory for market access in most global jurisdictions.

1.4.2.4. *Medical software lifecycle*

IEC 62304:2006 outlines lifecycle requirements for medical software, whether embedded within hardware or operating as standalone Software as a Medical Device (SaMD). It classifies software into three risk categories:

– *Class A*: No injury or damage possible.

– *Class B*: Non-serious injury possible.

– *Class C*: Death or serious injury possible.

Development, testing and documentation activities are scaled accordingly. With the rise of AI-driven diagnostics, telehealth platforms and mobile health apps, this standard ensures software reliability, security and traceability.

1.4.2.5. *Labeling and user information*

ISO 20417:2021 standardizes manufacturer-provided information, including labels, instructions for use (IFU) and technical documentation. It requires content to be clear, accessible and linguistically tailored to the end-user's region. By defining minimum labeling elements (e.g. device name, manufacturer details and UDI), structuring IFUs for usability and specifying safety warnings, this standard reduces use errors and aligns with the transparency mandates of regulations such as EU MDR.

1.4.2.6. *Cybersecurity management for medical devices: ISO/TS 20443*

This technical specification addresses the critical intersection of cybersecurity and patient safety in connected medical devices (ISO 13485). Unlike general information security standards, ISO/TS 20443 provides a tailored framework for the unique ecosystem of medical technology, focusing specifically on protecting the confidentiality, integrity and availability of health data while ensuring that security measures do not compromise device functionality or patient safety. It applies to both manufacturers developing connected devices (e.g. implantables, monitoring systems and SaMD) and healthcare organizations deploying these technologies. Key requirements include robust encryption protocols, vulnerability management processes, granular access controls and comprehensive incident response planning. As medical devices become increasingly interconnected and targeted by cyber threats, this standard offers a proactive approach to mitigating risks that could directly impact clinical operations and patient outcomes. Although still emerging, it is rapidly becoming a vital reference for complying with evolving regulatory expectations – including EU MDR cybersecurity annexes and FDA premarket guidance – and for building trust in digital health solutions.

At the hospital level, these requirements translate into strict procedures for technical acceptance, periodic servicing, decommissioning and materiovigilance. Regulatory audits also demand documented evidence of maintenance, tracking of technical alerts and corrective actions planes (European Commission 2017).

1.5. Managing uncertainty and random data

Maintenance of biomedical equipment presents a complex challenge due to the inherent uncertainty characterizing technical failures. Unlike some industrial sectors where failures often follow predictable patterns, medical devices are subject to numerous variable factors that make their maintenance particularly demanding. Usage intensity varies considerably across departments: a ventilator in intensive care operates continuously under maximum demand, while the same model in a standard care unit experiences more moderate use. This disparity results in radically different wear patterns that cannot be modeled using traditional approaches.

Environmental conditions represent another critical parameter. Excessive heat in sterilization rooms, humidity in operating theaters or electromagnetic interference in medical imaging create invisible but constant stresses on electronic and mechanical components. The same equipment may demonstrate completely different reliability, depending on its installation environment, challenging conventional preventive maintenance standards.

The human dimension is perhaps the most unpredictable factor. The considerable variation in technical skills among healthcare staff, combined with time pressures in clinical settings, creates demands that are sometimes unanticipated by manufacturers. Repeated improper handling during sensor connection, aggressive cleaning protocols or minor but frequent mechanical impacts can cause cumulative damage that standard protocols struggle to anticipate.

The advent of connected devices has introduced a new dimension of uncertainty: software interoperability. Updates to hospital information systems, conflicts between software versions or simply the complexity of interfaces between equipment from different manufacturers create systemic vulnerabilities. A failure may therefore stem not from a hardware defect, but from software incompatibility that emerges months after installation.

Faced with this complexity, biomedical maintenance departments must implement sophisticated strategies combining continuous IoT monitoring, predictive analytics using AI and adaptive user training programs. Maintenance is evolving from scheduled-based approaches to dynamic risk analysis methodologies, where the clinical criticality of each device determines the required supervision levels. Real-time data from embedded sensors enable detection of subtle anomalies long before they develop into full failures.

This transformation requires close collaboration between biomedical engineers, clinicians and information technology specialists. User feedback has become an essential source of intelligence for continuously improving prediction models. The objective is no longer to eliminate uncertainty – an impossible mission – but to develop organizational resilience to address it effectively, ensuring patient safety remains the absolute priority regardless of technological complexity.

1.5.1. *Sources of uncertainty in biomedical equipment maintenance*

Uncertainty in biomedical equipment maintenance arises from multiple interconnected factors that complicate prediction and planning efforts, creating a complex ecosystem of variables that challenge even the most sophisticated maintenance protocols.

– *Variability in failure patterns*

Identical device models may demonstrate significantly different failure characteristics and lifespans due to microscopic manufacturing variations, component sourcing differences and subtle production modifications over time. This stochastic behavior presents substantial challenges for accurate failure prediction, as traditional reliability models based on MTBF often prove inadequate for capturing the full spectrum of potential failure scenarios in medical devices.

– *Insufficient historical data*

Newly acquired devices, rarely used specialized equipment, or recently modified systems often lack comprehensive maintenance histories. This data scarcity impedes reliable statistical analysis and hinders the development of evidence-based maintenance strategies, particularly for predicting rare but critical failure modes that may only emerge after extended periods of operation or under specific clinical conditions.

– *Deviation from intended use conditions*

Clinical environments frequently deploy equipment in ways not anticipated by manufacturers, creating unexpected operational contexts that accelerate degradation. Devices may operate beyond recommended duty cycles, manage patient loads exceeding specifications or function in clinical scenarios not envisioned during the design phase, thereby accelerating wear and generating novel failure modes that bypass conventional diagnostic protocols.

– *Complex human–device interactions*

Many failures originate from use errors, inadequate training or workflow adaptations that introduce unanticipated stresses on equipment. These include improper calibration procedures, incorrect cleaning techniques, undocumented workarounds and accidental damage during emergency situations. Variability in operator skill levels and clinical experience further compounds these challenges, creating a layer of operational uncertainty that is difficult to quantify and mitigate through technical measures alone.

– *Environmental and cybersecurity threats*

Equipment performance is influenced by environmental conditions including temperature fluctuations, humidity levels, airborne contaminants and electromagnetic interference, which can degrade components and affect measurement accuracy. Furthermore, connected medical devices face growing vulnerabilities to malware infections, ransomware attacks and unauthorized access attempts that can compromise functionality and data integrity, introducing digital risk factors that traditional maintenance approaches are ill-equipped to address (Stawowy et al. 2021).

1.5.2. *Extended implications*

These uncertainty sources generate cascading effects on clinical operations that extend beyond immediate technical concerns. Unplanned downtime can disrupt patient care pathways, necessitate procedure rescheduling and increase dependence on backup equipment, potentially affecting clinical outcomes and patient safety. Financially, uncertainty leads to either excessive preventive maintenance (increasing operational costs and potentially causing unnecessary wear from frequent interventions) or insufficient maintenance (heightening failure risks and potentially leading to more severe consequences). Strategically, this environment demands adaptive maintenance approaches that balance resource allocation with clinical priorities while maintaining regulatory compliance across diverse equipment portfolios, requiring sophisticated risk assessment frameworks that can account for both technical and operational variables in healthcare settings.

1.6. Probabilistic methods

Probabilistic methods provide a mathematical framework to quantify uncertainties in biomedical equipment failures through statistical analysis of historical data. These approaches transform complex failure phenomena into predictive tools that optimize maintenance strategies, improve resource allocation and enable anticipation of critical failures. Their application spans all device categories, from basic electronic thermometers to advanced medical imaging systems, although their effectiveness varies significantly across different failure modes and device types.

1.6.1. *Weibull distribution*

The Weibull distribution models variable failure rates through its shape parameter β (indicating increasing, constant or decreasing failure rates) and scale parameter η (representing characteristic lifetime) (Chowdhuri et al. 2023). Particularly effective for mechanical and electromechanical components subject to wear mechanisms, it enables prediction of residual life and planning of targeted preventive maintenance. However, its application requires substantial failure data (typically exceeding 30 data points) for reliable parameter estimation and demonstrates limited effectiveness for failures originating from cyberattacks, human errors or software-related issues. The distribution's flexibility makes it valuable for modeling aging processes in infusion pumps, mechanical ventilators and surgical instruments.

1.6.2. *Exponential distribution*

As a particular case of the Weibull distribution (where $\beta = 1$), this model assumes constant failure rates over time. Mathematically straightforward to implement, it primarily suits electronic components and software systems where failures occur randomly and independent of time. While useful for preliminary modeling, its assumption of constant failure rate renders it inappropriate for devices exhibiting progressive wear patterns, and it may significantly underestimate risks when applied to equipment with time-dependent degradation mechanisms. The distribution finds appropriate application in solid-state electronics and embedded systems where failure mechanisms are truly random.

1.6.3. *Log-normal distribution*

This distribution models progressive degradation processes where the logarithm of failure times follows a normal distribution. It effectively represents cumulative wear in electrochemical components such as defibrillator batteries, physiological sensors and imaging system detectors.

The model provides realistic representations of gradual degradation but presents challenges in parameter interpretation and demonstrates sensitivity to preventive maintenance interventions. Its mathematical properties make it particularly suitable for modeling failure mechanisms dominated by diffusion processes or chemical degradation.

1.6.4. *Other distributions*

The gamma distribution generalizes the exponential distribution to model time until the *k*th failure in redundant systems, making it valuable for analyzing backup power systems and fault-tolerant architectures. The homogeneous Poisson process models failure counts over specified intervals with constant occurrence rates, proving useful for high-frequency failure events in patient monitoring equipment and mobile medical devices. More complex models, including Gaussian mixture distributions, address multifunctional devices with heterogeneous failure modes, although they require sophisticated parameter estimation techniques.

1.6.5. *AI integration and real-time data*

Contemporary approaches integrate traditional probabilistic methods with AI through several advanced methodologies. Machine learning algorithms, particularly survival analysis models and recurrent neural networks, incorporate multi-source operational data including usage patterns, environmental conditions and error logs to enhance prediction accuracy. Digital twin technology creates virtual replicas that simulate failure scenarios under varying operational conditions, enabling optimization of maintenance strategies through computational modeling. These advanced methods significantly improve failure prediction for complex systems but require substantial computational resources and specialized expertise for implementation.

1.6.6. *Limitations and future directions*

The effectiveness of probabilistic methods remains constrained by several fundamental factors. Data quality issues, including incomplete records, reporting inconsistencies and censored data, limit model accuracy. The dynamic nature of healthcare environments, including equipment repurposing and protocol changes, challenges the stability of model assumptions. Human factors and cyber-related failures escape conventional probabilistic modeling frameworks. Future research priorities include developing explainable AI systems for clinical engineering, establishing standardized reliability data exchange protocols between institutions and creating hybrid approaches that combine data-driven methodologies with physics-based models. These advancements aim to address current limitations while enhancing predictive capabilities for increasingly complex medical devices and systems.

1.7. Advanced predictive maintenance strategies for medical devices

1.7.1. *Analysis of reliability, maintainability and availability*

Statistical distributions are employed in the study of three key indicators to assess the performance of biomedical equipment.

– *Reliability*

Quantified by MTBF, reliability models the probability of fault-free operation of equipment over a specific period. The Weibull distribution, with its shape and scale parameters, is particularly useful for representing different failure patterns (infant mortality, useful life and progressive wear), thereby providing an essential predictive approach for strategic maintenance planning.

– *Maintainability*

Evaluated through MTTR (mean time to repair), maintainability integrates technical, logistical and human aspects of repair processes. It examines the effectiveness of corrective interventions, including spare parts availability, technical staff qualifications and post-repair validation procedures. Distributions such as log-normal effectively model the variability of intervention times, although this approach may underestimate the multifactorial complexity of failures in real clinical environments.

– *Availability*

These two dimensions are combined to calculate the effective operational rate (Availability = MTBF/(MTBF + MTTR) × 100%). This crucial indicator for care continuity provides an aggregated view that does not fully capture operational complexity, including administrative wait times, supply delays or organizational constraints.

1.7.2. *Advanced statistical tools*

– *Fault trees*

Deductively identify combinations of events leading to critical failures, proving indispensable for analyzing complex systems such as operating theaters or intensive care units. Their rigorous construction requires specialized expertise and a thorough understanding of system interactions.

Fault trees represent a systematic analysis method that deductively identifies combinations of events capable of causing critical failures. This approach proves particularly valuable when studying complex medical systems such as operating rooms or intensive care units, where equipment failure may lead to serious consequences.

The construction process follows top-down logic, beginning with a major adverse event and tracing backward to all potential causes through Boolean logic operators. The AND gate needs the simultaneous occurrence of all input events, while the OR gate indicates that a single input event suffices to trigger the output event.

In biomedical applications, this methodology enables analysis of critical scenarios such as intraoperative ventilation failure, while simultaneously considering hardware malfunctions, software errors, human factors and organizational deficiencies.

Rigorous implementation of this technique demands specialized expertise and profound understanding of system interactions. It additionally requires validation of assumptions by clinical experts and regular updates reflecting technological advancements.

Notable advantages include clear visualization of critical failure pathways and identification of single points of vulnerability. However, complexity escalates with system sophistication, and the method faces difficulties in fully capturing intricate human–system interactions.

Modern approaches now integrate fault trees with Monte Carlo simulations for probabilistic analysis and Bayesian networks for dynamic updating. Specialized software enables partial process automation, although human expertise remains essential for correct interpretation of results and validation of underlying assumptions.

– Markov chains

Model transitions between operational states (normal operation, partial failure, corrective maintenance and preventive maintenance) in dynamic systems, enabling the simulation of different maintenance strategies' impact on overall availability. Their application faces limitations in representing highly complex systems where interactions are nonlinear and multi-scale.

Markov chains provide a powerful mathematical framework for modeling transitions between distinct operational states – such as normal operation, partial failure, corrective maintenance and preventive maintenance – within dynamic systems. By representing these states and the probabilities of transitioning between them, Markov chains enable the simulation and analysis of how different maintenance strategies influence overall system availability, performance and longevity. This is especially valuable in environments like healthcare, where equipment reliability directly impacts patient outcomes. For instance, by modeling a ventilator's lifecycle – including periods of stable function, degradation, maintenance interventions and recovery – hospital engineers can quantitatively compare proactive versus reactive maintenance policies and optimize resource allocation.

However, the application of Markov chains faces significant limitations when dealing with highly complex systems characterized by nonlinear interactions and multi-scale dependencies. In real-world biomedical settings, equipment failure and repair processes often involve interdependencies that cannot be easily captured using Markovian assumptions, such as the requirement for memoryless transitions or fixed transition probabilities. For example, the failure rate of an MRI machine might depend not only on its usage patterns but also on environmental factors, software updates and human operational errors, which collectively introduce variability that challenges traditional Markov models. Additionally, systems with large numbers of states or continuous-state spaces may become computationally intractable when modeled using discrete-state Markov chains.

To address these shortcomings, modern approaches often integrate Markov chains with hybrid models or multi-scale modeling techniques. For instance, combining Markov decision processes (MDPs) with simulation-based methods allows for more flexible representation of complex systems, while still leveraging the interpretability and analytical power of Markov chains. Advances in computational tools have also facilitated the use of hierarchical and multi-level Markov models, which can partially accommodate nonlinearities by decomposing the system into simpler, interacting subsystems. Nevertheless, human expertise remains essential to properly structure these models, calibrate them with real-world data and interpret their outcomes in context-specific scenarios – ensuring that predictions align with practical constraints and operational realities.

– *Bayesian networks*

Bayesian networks integrate structured expert knowledge and historical data to dynamically update failure probabilities in real time. Particularly suited to uncertainty management and continuous learning, their implementation requires significant statistical expertise and often encounters cultural and technical barriers in healthcare environments.

Bayesian networks represent an advanced probabilistic methodology that dynamically integrates structured expert knowledge and historical data to update failure probabilities in real time. Unlike traditional statistical approaches, these networks enable graphical modeling of cause-and-effect relationships between technical, operational and human variables, providing a systemic view of risks. Their architecture relies on probabilistic theorems that quantify the impact of new observations on initial probabilities, creating a continuous learning system particularly suited to managing uncertainties in complex environments.

In the biomedical field, this approach proves crucial for predictive monitoring of critical equipment such as intensive care ventilators or medical imaging devices. By continuously assimilating maintenance data, technical alerts and usage feedback, Bayesian networks gradually refine their predictions and identify emerging failure patterns that conventional methods might overlook. For example, they can detect that specific combinations of high humidity levels and intensive usage cycles significantly increase failure risk in certain cardiac monitor models.

However, effective implementation of these networks faces several substantial challenges. The requirement for advanced statistical expertise to design and validate network structure often presents a major obstacle, particularly in hospital environments where specialized resources are limited. Furthermore, integrating data from heterogeneous sources (preventive maintenance records, clinical incident reports and technical parameters) requires complex preliminary standardization and robust data governance. Culturally, adoption of these tools can be hindered by reluctance to replace traditional methods with probabilistic approaches perceived as less transparent.

Despite these challenges, the advent of medical Industrial Internet of Things (IIoT) and improved software platforms are gradually facilitating deployment of these solutions. Modern Bayesian networks now incorporate machine learning capabilities that partially automate discovery of variable relationships, reducing initial dependence on comprehensive expert knowledge. Their strategic value lies in their ability to provide contextualized early warnings and support maintenance decisions based on multivariate and evolving risk understanding, ultimately contributing to enhanced patient safety and improved operational efficiency of biomedical services.

The integration of Bayesian networks with existing hospital information systems enables continuous adaptation to new clinical environments and technical constraints, representing a significant advancement in predictive maintenance strategies for medical technology management.

1.7.3. *Operational applications*

The results of these quantitative analyses guide crucial management decisions: scheduled replacement of equipment at end of economic life, optimized planning of preventive revisions based on actual device condition and strategic evaluation of cost–benefit ratios for technological upgrades.

1.7.4. *Real-time data integration*

The advent of the biomedical IIoT enables dynamic calculation of MTBF/MTTR indicators via connected sensors, early detection of reliability deviations through continuous operational data analysis and real-time adaptation of statistical models to actual usage conditions. This technological evolution nevertheless presents substantial challenges in massive data processing and cybersecurity of connected medical devices.

1.7.5. *Limitations and best practices*

Several limitations persist: maintenance data heterogeneity frequently biases historical analyses, the dual expertise required in advanced statistics and biomedical engineering remains scarce in the job market and necessary contextualization of models to clinical specificities is often underestimated in practical implementations.

1.7.6. *Perspectives*

The future of these predictive analyses lies in the synergistic integration of AI (advanced machine learning) and digital twins. These emerging technologies enable more accurate forecasting and genuinely predictive, personalized maintenance, provided that interoperable data standards are developed and biomedical maintenance teams' skills are significantly enhanced. The progressive maturation of evidence-based maintenance approaches should enable refined resource optimization while ensuring patient safety and sustainability of technological investments.

1.8. Integrating digitalization and AI into hospital maintenance

The emergence of digital technologies is fundamentally transforming medical equipment management by enabling a proactive, data-driven approach. Through the Internet of Things (IoT), medical devices are becoming smart and connected, continuously transmitting operational data such as operating parameters, critical temperatures, load cycles, abnormal vibrations and system errors. These data streams feed CBM models, where interventions are triggered only in response to measurable anomalies, thereby eliminating unnecessary maintenance while preventing unexpected failures.

More sophisticated predictive maintenance (PdM) utilizes AI to anticipate potential failures before they occur (Alkhatib et al. 2025) by analyzing complex patterns and hidden correlations within both historical and real-time data.

Advanced algorithms process these multi-source data, cross-reference maintenance histories, detect subtle deviations and estimate failure probabilities with increasing accuracy. These systems employ machine learning for trend analysis, deep learning for recognizing complex patterns in high-dimensional data and neural networks to model nonlinear relationships between input variables and failure risks.

1.8.1. *Benefits and implications*

– *Reduction of unplanned downtime*

Anticipating critical failures enables scheduling of interventions during planned maintenance windows, avoiding disruptions during urgent clinical procedures.

– *Optimization of maintenance schedules*

Collected data allow for prioritizing interventions based on the actual condition of equipment and dynamically adjusting the allocation of technical and human resources.

– *Extended equipment lifespan*

Targeted maintenance reduces unnecessary stress and excessive preventive disassembly, preserving the mechanical and functional integrity of devices.

– *Enhanced patient safety*

Reducing unexpected failures during critical medical procedures (e.g. surgery and intensive care) directly mitigates risks to patients.

1.8.2. *Challenges and considerations*

– *Data quality and interoperability*

Prediction reliability depends on the completeness, accuracy and standardization of data from multiple sources (sensors, digital multimeters and computerized maintenance management systems).

– *Specialized skills*

Designing, deploying and maintaining these systems requires cross-disciplinary expertise in biomedical engineering, data science and computer science.

– *Cybersecurity*

Increasing connectivity of medical devices expands the potential attack surface, necessitating robust security protocols to protect both data and critical equipment functions.

– *Cultural change*

Transitioning from calendar-based to data-driven maintenance involves transforming established processes and securing buy-in from technical and clinical teams.

As these technologies mature, they are redefining the paradigms of technical and clinical management of biomedical equipment, paving the way for truly predictive maintenance integrated into the continuum of care.

1.9. Economic, environmental and strategic challenges in medical equipment management

Effective medical equipment management needs a comprehensive economic approach that extends far beyond initial acquisition costs. The global cost of ownership (TCO) encompasses numerous factors including logistics and installation, consumables and reagents, preventive and corrective maintenance, technical training and certification, software updates and cybersecurity, sterilization and disinfection procedures, and eventual end-of-life disposal or recycling.

This holistic financial perspective reveals that operational expenses typically exceed purchase prices by three to five times over a device's lifecycle.

Contemporary management trends increasingly emphasize sustainable and socially responsible practices through several innovative strategies:

– *Equipment refurbishment*

Advanced technical processes restore used medical devices to original equipment manufacturer (OEM) specifications, typically at 40%–60% of the cost of new equipment while maintaining equivalent performance and safety standards.

– *Inter-hospital resource sharing*

Strategic pooling of high-value specialized equipment through regional networks or consortium agreements maximizes utilization rates while distributing operational costs across multiple institutions.

– *Circular economy implementation*

Comprehensive recycling programs recover valuable materials (precious metals, rare earth elements and plastics) from decommissioned equipment, while proper hazardous material handling ensures environmental compliance.

– *Eco-design integration*

Preferred supplier selection based on environmental criteria including energy efficiency, reduced material footprint, design for disassembly and minimized use of hazardous substances throughout the product lifecycle (Mengual et al. 2014).

From a strategic perspective, effective biomedical fleet management directly contributes to enhanced care quality through improved equipment availability and reliability, increased patient satisfaction through uninterrupted service delivery, ensured regulatory compliance with evolving environmental and safety standards, and strengthened institutional reputation as a healthcare sustainability leader.

In an increasingly competitive healthcare landscape, the ability to maintain a technologically advanced, environmentally responsible and financially sustainable equipment ecosystem has emerged as a significant competitive differentiator. Institutions that successfully implement integrated lifecycle management strategies not only reduce operational costs but also demonstrate organizational excellence and commitment to both patient care and environmental stewardship, ultimately positioning themselves as preferred providers in value-based healthcare systems.

1.10. Conclusion

Medical equipment has become a cornerstone of modern healthcare facilities, with its performance directly determining clinical efficiency, patient safety and organizational sustainability. Confronted with complex technical, regulatory, economic and human challenges, an integrated approach has become essential. This approach combines advanced technological tools such as AI and IoT, rigorous lifecycle equipment management, an organizational culture focused on safety and continuous anticipation of emerging risks. Biomedical maintenance can no longer be perceived as merely a support function – it represents a central strategic element for the success of 21st-century healthcare institutions, requiring a cross-functional vision that integrates digital transformation, environmental responsibility and resource optimization while ensuring excellence in care. Its evolution toward predictive, data-driven maintenance makes it an essential lever for reconciling technological innovation, economic constraints and health safety imperatives, thereby positioning biomedical services as key players in overall healthcare system performance within value-based healthcare models.

2

Description of the Dialysis Machine

2.1. Introduction

The kidneys are vital, life-sustaining organs responsible for purifying the blood of metabolic toxins and regulating fluid and electrolyte homeostasis. End-stage renal disease (ESRD) results in the accumulation of harmful waste products, such as urea and creatinine, leading to severe and potentially life-threatening systemic complications affecting the cardiovascular, neurological and hematological systems.

To address the loss of kidney function, three principal renal replacement therapies are available: hemodialysis, peritoneal dialysis and kidney transplantation. Tunisia has progressively developed its renal care infrastructure, initiating a chronic hemodialysis program in 1968, introducing peritoneal dialysis in 1982 and performing its first kidney transplant in 1986 (Fidanoglu et al. 2017). Hemodialysis remains the most prevalent treatment modality both in Tunisia and globally (Bahreini et al. 2018), a fact underscored by the significant expansion of its infrastructure: the number of centers grew from 10 in 1986 to 168 in 2019, comprising a total of 3,014 machines according to the Tunisian Ministry of Health (Mazah 2019).

Consequently, the reliability of hemodialysis machines is paramount for nephrologists and clinical engineers. These devices are highly sophisticated systems, not mere pumps; they perform critical functions including blood

---

For a color version of all the figures in this chapter, see www.iste.co.uk/bouchoucha/optimizing.zip.

filtration, precise dialysate preparation (controlling temperature, conductivity and pH) and continuous monitoring of patient safety (detecting air, blood leaks and pressure anomalies). Their flawless operation is essential to ensure immediate patient safety – preventing incidents such as hypotensive episodes or air embolisms – as well as the long-term efficacy and continuity of treatment.

The correlation between maintenance quality and clinical outcomes is well-substantiated. A study analyzing 134 patients on a thrice-weekly dialysis regimen across 48 machines over nine years demonstrated that improper maintenance directly leads to inadequate dialysis sessions (reflected by a reduced dose, or inadequate Kt/V) and a higher rate of complications, including anemia, hyperphosphatemia and increased hospitalization rates (Azar 2009). Furthermore, machine failures cause stressful session interruptions and significant operational disruption for dialysis centers.

Therefore, implementing a proactive and effective maintenance strategy is imperative. There is a critical need to develop practical, efficient and standardized methodologies that can be adopted by healthcare facility maintenance departments to optimize the availability and reliability of hemodialysis equipment. The ultimate goal is to safeguard absolute patient safety and ensure the therapeutic efficacy of every dialysis treatment, thereby directly improving the quality of care and long-term prognosis for patients with kidney failure.

2.2. Functions and operational principles of a hemodialysis machine

The dialysis machine is sometimes referred to as "the artificial kidney", but this is not really correct, because it is in the dialyzer that the blood purification takes place. Still, it is true that the dialysis machine is needed to perform hemodialysis.

Hemodialysis machines represent a critical category of medical devices directly linked to patient survival, particularly in the management of kidney failure. Acting as an artificial kidney, the hemodialysis system circulates both the patient's blood and a cleansing solution known as dialysate through a specialized filter called a dialyzer. The dialysate, composed of water, electrolytes and salts, plays a vital role in purifying the blood by eliminating metabolic waste and maintaining electrolyte balance within the filter. A semi-permeable membrane separates the blood from the dialysate, allowing selective transfer of solutes and fluids according to physical diffusion and osmosis principles (Madhukar 2005).

Thus, while the dialyzer is aptly nicknamed the "artificial kidney" for its filtration role, the dialysis machine functions as its indispensable life-sustaining control system. Far more than a simple pump, it is a highly integrated system of precision and safety. It meticulously regulates every critical treatment parameter: it pumps blood at a precisely controlled flow rate, prepares and heats dialysate to a rigorous standard while continuously monitoring its composition (via conductivity) and oversees the entire procedure through a suite of sensors and monitors. These systems provide real-time surveillance of arterial and venous pressures, detect the presence of air in the extracorporeal circuit to prevent fatal gas embolism and identify potential blood or dialysate leaks. Consequently, the machine is the element that orchestrates and safeguards the environment in which the dialyzer performs its purification, making their synergistic combination the true "artificial kidney" in a clinical setting.

To understand the hemodialysis purification process, it is conceptually divided into two main interdependent circuits: the blood circuit and the dialysate circuit, as illustrated in Figure 2.1. This fundamental schematic framework allows for a clear conceptualization of hemodialysis as a critically controlled interaction between the patient's blood and the dialysate – an interaction that is central to the detoxification process.

The blood circuit: is an extracorporeal pathway that transports the patient's blood to the dialyzer. Propelled by a peristaltic blood pump, blood leaves the body through a vascular access point, circulates through the hollow fibers of the dialyzer where solute exchange occurs and is subsequently returned to the body, purified. This circuit is continuously monitored by pressure sensors and air detectors to ensure the integrity and safety of the blood throughout the procedure.

The dialysate circuit: is responsible for preparing and circulating the dialysate, a sterile solution of water, electrolytes and salts with a precisely calibrated composition. This fluid flows counter-currently around the outside of the hollow fibers, creating and maintaining the concentration gradients essential for removing uremic toxins and normalizing electrolyte levels through the process of diffusion.

A semi-permeable membrane within the dialyzer serves as the critical physicochemical interface, allowing these two circuits to interact without direct contact. Blood purification occurs across this membrane via the tightly regulated mechanisms of diffusion and ultrafiltration (UF), processes that are meticulously controlled and monitored by the dialysis machine to ensure both efficacy and patient safety.

A hemodialysis machine is a sophisticated life-support system designed to perform the blood-cleansing functions of failed kidneys artificially. Its operation is a complex interplay of mechanics, chemistry and patient safety protocols. The machine's functions can be comprehensively divided into the following categories.

2.2.1. *Basic functions: the core dialysis process*

These functions form the essential therapeutic action of hemodialysis, facilitating the exchange of toxins and excess fluids between the patient's blood and the dialysate across a semi-permeable membrane in the dialyzer (Figure 2.2).

2.2.1.1. *Working principle*

The dialyzer, or artificial kidney, purifies the blood of patients suffering from renal failure by replicating the kidney's filtration functions. Its principle is based on the exchange of substances across a semi-permeable membrane, separating the blood from the dialysate, through the processes of diffusion and UF.

The patient's blood is drawn via vascular access and circulates through the blood compartment, consisting of thousands of hollow fibers. The semi-permeable membrane, which forms the wall of these fibers, is the core element. Its microscopic pores allow uremic toxins (such as urea and creatinine) and excess electrolytes to pass through, while retaining blood cells and proteins.

The dialysate, a sterile solution circulating around the fibers in the external compartment, plays an active role. Its ionic composition is similar to that of healthy plasma, allowing, by diffusion, excess solutes in the blood to migrate into the dialysate where their concentration is zero or low. Conversely, substances such as bicarbonate can diffuse into the blood to correct acidosis. UF is controlled by a negative pressure applied to the dialysate, forcing the removal of excess water from the blood. Finally, the headers of the dialyzer ensure a homogeneous distribution of fluids in their respective compartments to optimize the overall efficiency of the dialysis process.

The core process relies on two key physical principles as follows (Figure 2.2):

– *Diffusion*

The transfer of solutes such as urea, creatinine, potassium and hydrogen ions occurs by diffusion, moving from the patient's blood, where their concentration is higher, toward the dialysate, where their concentration is lower, across the pores of the dialyzer's semi-permeable membrane. Conversely, bicarbonate ions diffuse in the opposite direction – from the dialysate into the bloodstream – in order to help correct metabolic acidosis.

– *Ultrafiltration*

The extraction of excess fluid during hemodialysis is achieved by creating a pressure gradient across the semi-permeable membrane, typically through the application of negative pressure on the dialysate side. This gradient drives the movement of water out of the bloodstream, carrying with it small solutes capable of passing through the membrane's pores. In this way, the process effectively alleviates fluid overload and restores the patient's fluid balance.

2.2.1.2. *Key components and their roles*

2.2.1.2.1. The extracorporeal blood circuit

Blood pump: a peristaltic pump that acts as the heart of the extracorporeal circuit. It propels blood from the patient through the dialysis tubing and dialyzer at a precisely controlled flow rate (typically 250–450 mL/min), which is a primary determinant of treatment efficacy.

Vascular access: the connection point to the patient's bloodstream either via an arteriovenous fistula/graft or a central venous catheter.

Arterial and venous pressure monitors: these sensors monitor pressure before (arterial) and after (venous) the blood pump. Sudden changes can indicate problems such as a kink in the bloodline, a clotting access or a needle dislodgement.

Air bubble detector (air trap): located on the venous bloodline, this ultrasonic sensor detects the presence of air bubbles. Upon detection, it triggers an alarm and immediately clamps the venous line to prevent a potentially fatal air embolism from returning to the patient.

2.2.1.2.2. The dialysate circuit

Water treatment system: the foundation of safe dialysis. City water is pre-treated through sediment filters, carbon tanks and finally a reverse osmosis (RO) system. The RO membrane removes over 90% of contaminants, ions and bacteria, producing highly pure water essential for dialysate preparation.

Proportioning system (mixing unit): the core of the dialysate circuit. It accurately blends the purified water with concentrated electrolyte solutions (acid concentrate and bicarbonate concentrate) at a specific ratio (e.g. 1 part concentrate to 34 parts water) to create the final dialysate with the prescribed chemical composition.

Conductivity sensor: continuously measures the ionic concentration of the prepared dialysate. If the conductivity falls outside strict limits, it indicates improper mixing, and the machine will divert (bypass) the non-compliant fluid away from the dialyzer and sound an alarm.

Temperature sensor: heats the dialysate to body temperature (usually 36–37°C). This prevents patient discomfort and protects blood cells from thermal damage (hemolysis).

Dialysate pump: circulates the dialysate through the dialyzer at a controlled flow rate (usually 500–800 mL/min), often in a countercurrent direction to the blood flow to maximize the efficiency of toxin removal.

2.2.2. *Safety functions: the automated vigilance system*

2.2.2.1. *Principle*

Patient safety is paramount. The machine incorporates a multi-layered, redundant system of monitors and alarms that constantly supervise the treatment and can automatically intervene.

Continuous monitoring of all critical parameters with automatic responses (alarms, bypass and circuit clamping) if preset safety limits are exceeded.

2.2.2.2. *Key safety components*

– *Blood leak detector*: positioned on the effluent dialysate line after the dialyzer, it uses optical sensors to detect even minute traces of blood, which would indicate a rupture in the dialyzer membrane. It triggers a high-priority alarm and stops treatment.

– *Venous line clamp*: an electrically operated clamp that automatically engages on the venous bloodline during any critical alarm (e.g. air detection, high venous pressure, and blood leak), physically preventing blood return to the patient until the problem is resolved.

– *Monitoring of pressure and ultrafiltration*: the machine continuously calculates the transmembrane pressure (TMP), the driving force for fluid removal. Sophisticated volumetric or flow-based systems ensure that the exact prescribed amount of fluid (UF goal) is removed, preventing hypotension or fluid overload.

2.2.3. *Disinfection circuit: ensuring microbiological safety*

After each treatment, the machine's internal fluid pathways must be thoroughly cleaned and disinfected to prevent bacterial growth (biofilm) and cross-contamination between patients.

– *Principle*: it includes circulating hot water and chemical disinfectants (e.g. peracetic acid, sodium hypochlorite) through a closed-loop system that includes all parts of the dialysate circuit.

– *Procedure*: an automated disinfection program flushes saline and water through the system, then heats and circulates the disinfectant, and finally performs a rinse cycle to eliminate any chemical residues. This ensures the machine is sterile and safe for the next patient.

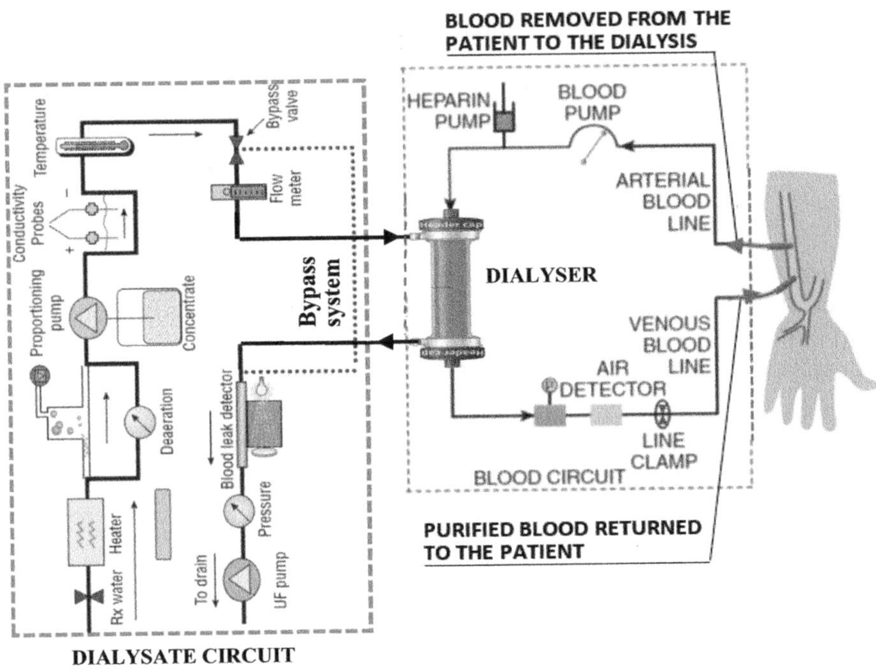

Figure 2.1. *Hemodialysis system components*

The risks of hemodialysis extend beyond the machine itself and require a holistic safety culture, as this is the leading cause of morbidity in dialysis patients. It is mitigated by a three-pillar approach:

– *Water quality*: imperfect water purification can lead to pyrogenic reactions (fever and chills) from bacterial endotoxins or chemical intoxication (e.g. from chloramines).

– *Rigorous disinfection protocols*: these include strict procedures for both the machine and the dialysis station environment (surface cleaning, handling of supplies and waste disposal).

– *Aseptic technique*: meticulous hand hygiene and sterile technique when handling the patient's vascular access are critical to prevent bloodstream infections.

– *Preventive maintenance and calibration*: the reliability of every sensor, pump and alarm is non-negotiable. Regular scheduled maintenance and calibration are mandatory. A malfunctioning conductivity sensor or an uncalibrated pressure monitor can lead to serious, undetected patient harm.

In summary, a modern hemodialysis machine is far more than a simple pump; it is an integrated life-support system. Its therapeutic effectiveness is built upon the precise execution of its basic functions, which is itself entirely dependent on a robust, multi-faceted safety infrastructure and rigorous hygiene and maintenance protocols.

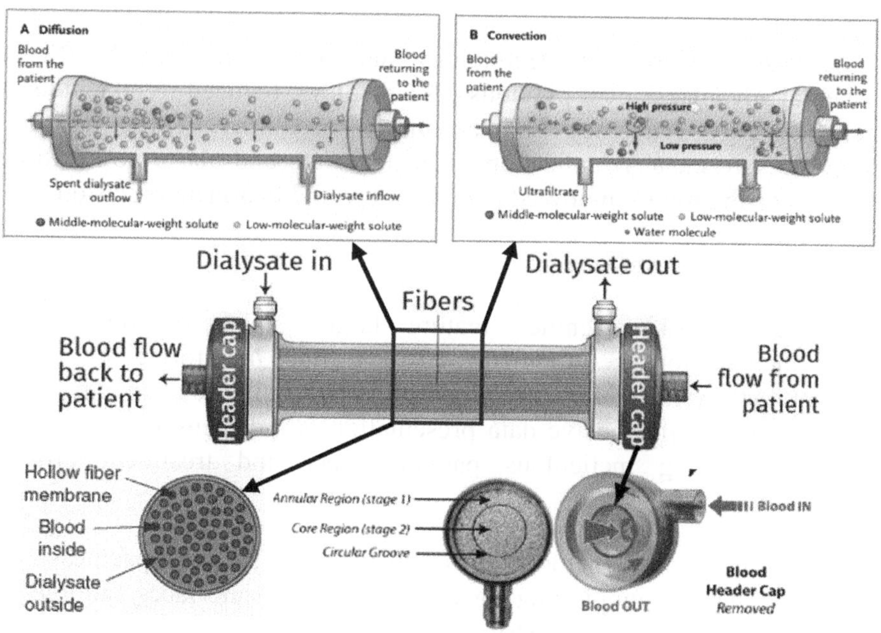

Figure 2.2. *The dialyzer*

2.3. Dialysis machine features

2.3.1. *Operator controls*

– Display and backlight

Clinical dialysis machines are equipped with large displays (typically 15–24 inches) that enable the simultaneous monitoring of multiple treatment parameters in real-time: blood pressure, dialysate conductivity, UF rate, TMP and treatment duration. This comprehensive interface is essential for healthcare providers to continuously assess patient safety and treatment efficacy.

In contrast, home hemodialysis machines prioritize compactness and user-friendliness, featuring smaller screens and simplified interfaces designed for individual operation. The evolution of user interface technology has been transformative: modern graphical user interfaces (GUIs) have largely replaced earlier button-based systems, significantly reducing cognitive load and minimizing potential operational errors.

The transition to LCD technology has been instrumental, offering substantial advantages over previous CRT displays: reduced weight and size, superior energy efficiency and enhanced readability. These advanced displays ensure clear visualization of critical data and alarm notifications, directly contributing to improved patient safety and treatment precision.

– Touch screens and keyboards

Clinical dialysis machines feature large, high-resolution displays (15–24 inches) for the simultaneous, real-time monitoring of critical treatment parameters – including blood pressure, dialysate conductivity and UF rate. This comprehensive data presentation is indispensable for clinical oversight, enabling meticulous patient safety and treatment efficacy management.

Conversely, home hemodialysis units prioritize compactness and user-centric design, employing smaller, simplified interfaces tailored for individual operation. The industry-wide shift to intuitive GUIs has markedly reduced cognitive load and minimized error. This evolution, coupled with the superior energy efficiency, reduced weight and enhanced clarity of LCD technology, ensures precise data and alert visualization, directly contributing to treatment accuracy and paramount patient safety.

2.3.2. *Patient connections*

A fundamental principle of hemodialysis is that the core purification process involves no direct electrical contact with the patient, ensuring intrinsic safety from electrical hazards. However, advancing technology has led to the integration of sophisticated optional modules in many modern dialysis systems. These modules non-invasively monitor a suite of vital signs, including heart rate, continuous blood pressure, respiration rate and core temperature, delivering a holistic overview of the patient's physiological status throughout the procedure. While these enhanced monitoring capabilities significantly augment clinical oversight and patient safety, they are ancillary features rather than essential to the central dialytic function of toxin removal and fluid management. Consequently, this application note will focus exclusively on the essential components and processes specific to dialysis itself.

2.3.3. *Data interfaces*

A comprehensive digital log of every dialysis session is automatically generated and electronically archived, creating a detailed and immutable record for each patient. These data are accessible through multiple integrated connectivity options designed for seamless interoperability with healthcare infrastructure. Modern dialysis machines are equipped with a suite of physical data ports, including USB for local access, Ethernet for wired network integration and various serial interfaces (RS-232, RS-485 and RS-422) to ensure compatibility with legacy hospital information systems. Furthermore, contemporary units often feature advanced wireless interfaces like Wi-Fi, enabling secure, real-time data transmission directly to electronic medical records (EMR) and cloud-based platforms. This robust connectivity architecture supports remote monitoring, streamlined data management and enhances both the continuity of care and the potential for data-driven clinical analysis, while maintaining strict cybersecurity and patient privacy protocols.

2.3.4. *Mechanism*

– *Pumps*

Peristaltic pumps are widely employed in hemodialysis machines to circulate large volumes of fluids such as blood, dialysate, water and saline. Their main advantage lies in the fact that the pumped liquid never comes into direct contact with the mechanical components of the device. Instead, the fluid flows through a flexible tube that is periodically compressed by rotating rollers, creating forward motion. These pumps consume considerable energy and are typically powered by DC or AC motors equipped with variable speed regulation. To maintain accurate flow rates, electronic control systems are used to monitor and adjust the motor's rotational speed.

– *Valves*

Multiple electronically actuated valves are incorporated into the machine to enable adjustable mixing ratios. These valves can range from basic on/off types operated by solenoids to highly precise variable-position valves controlled by stepper motors or alternative mechanisms, depending on the level of control required.

– *Sensors*

Key parameters under constant surveillance include blood pressure at multiple points (pre-pump arterial, pre-dialyzer and venous) to detect access complications or line obstructions, dialysate pressure and temperature to guarantee the integrity and comfort of the treatment, and conductivity to verify the correct composition of the dialysate fluid. Furthermore, blood leak detectors scrutinize the effluent dialysate for the presence of hemoglobin, indicating a membrane rupture. Most critically, air bubble detectors in the venous line stand as the last line of defense against a potentially fatal air embolism. The TMP, calculated from the blood and dialysate pressure gradients, is precisely regulated to control the rate of fluid removal (UF). This network of sensors allows the machine's processors to maintain equilibrium and respond instantaneously to any deviation, triggering alarms and automatic shut-down procedures to protect the patient.

– *Cleaning system*

To prevent biofilm formation and cross-contamination between patients, a rigorous disinfection and sterilization protocol is mandatory after each dialysis session. This process targets all internal fluid pathways, including both the dialysate circuit and the extracorporeal blood circuit, which must be cleansed of all biological residues and chemical contaminants.

One prevalent method is thermal disinfection, which involves heating ultrapure water to a high sterilizing temperature (often above 80–95°C) and circulating it through the entire hydraulic system for a prescribed duration. This effectively pasteurizes the internal components. Alternatively, or in conjunction with heat, chemical disinfection is employed using agents like citric acid, peracetic acid or sodium hypochlorite to eradicate microbes and dissolve mineral deposits.

Modern dialysis machines feature fully automated disinfection cycles programmed by the manufacturer. These cycles require precise monitoring, as the machine's internal computers must regulate temperature, concentration of chemical agents, contact time, flow rates and finally a thorough rinsing cycle to ensure no disinfectant residue remains. This automated, validated process is critical for operational readiness and, most importantly, for ensuring patient safety against infection and exposure to harmful residues.

2.3.5. *Processing*

– *Microcontrollers*

The operational complexity of a modern dialysis machine necessitates a distributed computational architecture. Rather than relying on a single central processor, these systems employ numerous dedicated microcontrollers, each acting as an independent, specialized brain for a specific subsystem.

For instance, one microcontroller may be solely responsible for the precise closed-loop control of the blood pump motor, ensuring accurate speed and flow rate. Another might dedicate its processing power to continuously sampling data from a suite of sensors – such as pressure transducers, conductivity cells and air detectors – in real-time. This decentralized approach enhances both system reliability and responsiveness. By offloading tasks to dedicated processors, the machine can execute critical safety checks and control functions simultaneously and without interruption, forming a robust, fail-safe network that is fundamental to protecting patient safety and ensuring consistent treatment delivery. This architecture also allows for more sophisticated fault detection and simplified maintenance, as issues can often be isolated to a specific functional module.

– *Fail-safe circuitry*

Patient safety is paramount, and dialysis machines are designed with a multi-layered, fault-tolerant architecture. This incorporates specialized integrated circuits (ICs) and microprocessors equipped with advanced self-test (built-in self-test [BIST]) and real-time fault-reporting capabilities. These systems continuously monitor their own integrity and the validity of critical signals, ensuring the machine can maintain a safe state even in the event of a single internal component failure.

Upon detecting a deviation – whether from a sensor reading exceeding safe limits or an internal hardware fault – the circuitry immediately initiates a hierarchical response. This always includes triggering both audible alarms (of varying urgency and pattern) and prominent visual alerts on the user interface, often accompanied by specific error codes. Crucially, the response also involves automatic physical safety interventions, such as halting pumps and engaging venous line clamps, to immediately isolate the patient from potential harm. This robust fail-safe design ensures that no single point of failure can compromise patient safety.

2.3.6. *Power supply*

The extended and critical nature of hemodialysis treatments mandates that all dialysis equipment is directly powered by a reliable AC mains connection, ensuring uninterrupted operation throughout a session that can last several hours. The power supply architecture is therefore engineered to the highest standards of medical-grade safety and reliability.

At its core, the system employs specialized AC-DC converters that comply with stringent medical safety standards (such as IEC 60601-1), providing exceptional electrical isolation to protect the patient from any risk of leakage current or macro-shock. Internally, the power management system is complex: high-efficiency switching regulators generate the multiple voltage rails required for motors, pumps and digital logic. However, for the machine's most noise-sensitive precision circuits – such as analog sensors measuring conductivity or pressure – these switching outputs are further stabilized by ultra-low-noise linear regulators situated close to the point of load. This hybrid approach ensures signal integrity and measurement accuracy.

Furthermore, adhering to safety regulations, the power supply incorporates continuous self-monitoring systems that vigilantly track its own operational status, including input/output voltage levels, temperature and current flow. Any anomaly, such as an overvoltage, overtemperature or overload condition, triggers immediate protective actions and system alarms to safeguard both the patient and the integrity of the ongoing treatment.

2.4. Maintenance strategies for hemodialysis machines: types, components and a real case study

Hemodialysis machines are complex biomedical systems requiring rigorous and structured maintenance strategies to ensure patient safety and treatment efficacy. These strategies fall into four main categories: preventive, corrective, predictive and condition-based maintenance, each targeting specific components depending on their critical function, failure risk and performance degradation profiles.

2.4.1. *Preventive maintenance*

Consists of planned interventions performed at regular intervals to minimize wear and avoid failure. It typically involves routine disinfection, calibration of conductivity and temperature sensors and inspection of pressure transducers, air traps and heparin pumps. For instance, most manufacturers recommend monthly disinfection of the dialysate circuit and calibration of the proportional valve controlling dialysate mixing to avoid conductivity deviations (Saleh et al. 2014). Neglecting preventive maintenance on the UF pump has been associated with uncontrolled fluid removal and hypotension events during treatment sessions (Wang et al. 2024).

2.4.2. *Corrective maintenance*

Addresses failures after they occur. This approach is often urgent, especially when critical safety components like the blood leak detector, air bubble sensor or venous pressure monitor malfunction. A real-life case was reported in a German dialysis unit, where a blood leak detector failed during a session, leading to contamination of the dialysate circuit. The patient experienced mild hemolysis, and the machine was immediately shut down for emergency repair (Tsarouhas 2012). The root cause was traced to a faulty infrared sensor, which had not been replaced despite several alerts logged in the system.

2.4.3. *Predictive maintenance*

Aims to prevent failure using real-time data trends and machine learning algorithms to forecast degradation. Commonly monitored components include blood pump motors, solenoid valves and cooling fans. Techniques such as vibration analysis, thermal profiling and resistance monitoring are used to assess component health. In a Spanish clinic, continuous monitoring of the peristaltic pump's motor current detected subtle anomalies, prompting early replacement and preventing sudden failure during a scheduled dialysis session (Namuduri et al. 2020).

2.4.4. *Condition-based maintenance*

Relies on threshold-triggered interventions based on performance deviations. When operational parameters like TMP or dialysate flow rate exceed predefined limits, specific components – such as TMP sensors, clogged dialyzers or partially occluded tubing – are examined and maintained. A notable example involved a rise in TMP over several sessions at a French facility. Maintenance staff discovered mineral scaling in the bicarbonate mixing chamber, which was cleaned and retrofitted with improved filters to prevent recurrence (Mutia et al. 2018).

In conclusion, each maintenance type plays a complementary role in sustaining the safe operation of hemodialysis machines. Preventive and predictive maintenance enhance reliability and minimize treatment interruption, while corrective and condition-based actions ensure rapid response to detected issues. Integrating these strategies with detailed component monitoring enables proactive interventions, reduces patient risk and optimizes machine lifespan.

2.5. Presentation of the studied dialysis machines

Our study will focus on a park of 10 dialysis machines located at the Grombalia hospital under the regional health department of Nabeul, Tunisian Ministry of Health. It should be noted that all machines have the same mark Fresinus.

The Fresinus hemodialysis machine is engineered to perform dialysis procedures in hospitals, specialized clinics or at home under the supervision of a trained operator other than the patient. It is suitable for individuals with either chronic or acute kidney failure.

2.5.1. *Characteristics of the Fresinus hemodialysis machine*

The Fresinus hemodialysis device is engineered to administer dialysis treatment by accurately regulating and observing both the dialysate and extracorporeal blood pathways. In the extracorporeal blood circuit, blood is continuously extracted from the patient and directed through a dialyzer, where a semi-permeable membrane removes metabolic waste before the cleansed blood is returned to the patient.

During the procedure, the system continuously tracks arterial and venous pressures and detects air bubbles or any blood leakage. Additionally, the machine enables precise heparin delivery to ensure proper anticoagulation throughout the dialysis session.

In the dialysate circuit, concentrates are combined with purified water, then heated, degassed and delivered to the dialyzer. Balancing chambers ensure that the volume of dialysate entering the dialyzer matches the volume exiting it, thereby regulating UF and maintaining fluid balance for the patient.

Figures 2.3 and 2.4 and Table 2.1 illustrate the front and rear view of the Fresinus hemodialysis machine.

2.5.2. *Overview of dialysis machine failure history*

The considered dialysis system is composed of 10 hemodialysis machines (M1, M2, . . ., M10), Fresinus mark and 40088 type. Their commissioning date is August 11, 2012. The dialysis system operates with an average about of 3.5 h per working day, 6 days a week and usually stops on Sundays.

The failure history of the 10 machines was recorded for the period from 2013 to 2022 by the hospital's technical department. The data contain the detected failures and their nature, the corrective actions taken to overcome the fault, the downtime and the exact time of the breakdown.

The time between failure (TBF) of the component is given as the period from the moment when the equipment starts functioning after a failure/maintenance, until the moment it fails again, while the time to repair (TTR) a damaged component is given as the period from the moment of failure to the moment it begins functioning again. Both TTR and TBF are calculated and saved in hours for 10 machines.

The maintenance policy currently applied to the dialysis machines is both preventive and corrective. Preventive maintenance is planned and applied periodically, while corrective maintenance is unplanned and performed whenever a fault is detected, the procedure needs emergency intervention by maintenance personnel to return the device to operating condition. Preventive maintenance is performed every 2,500 h of operation.

Appendix 1 illustrates the failure history of all machines, the date and the nature of the failure are given as well as the TBF and TTR, which are calculated for each intervention.

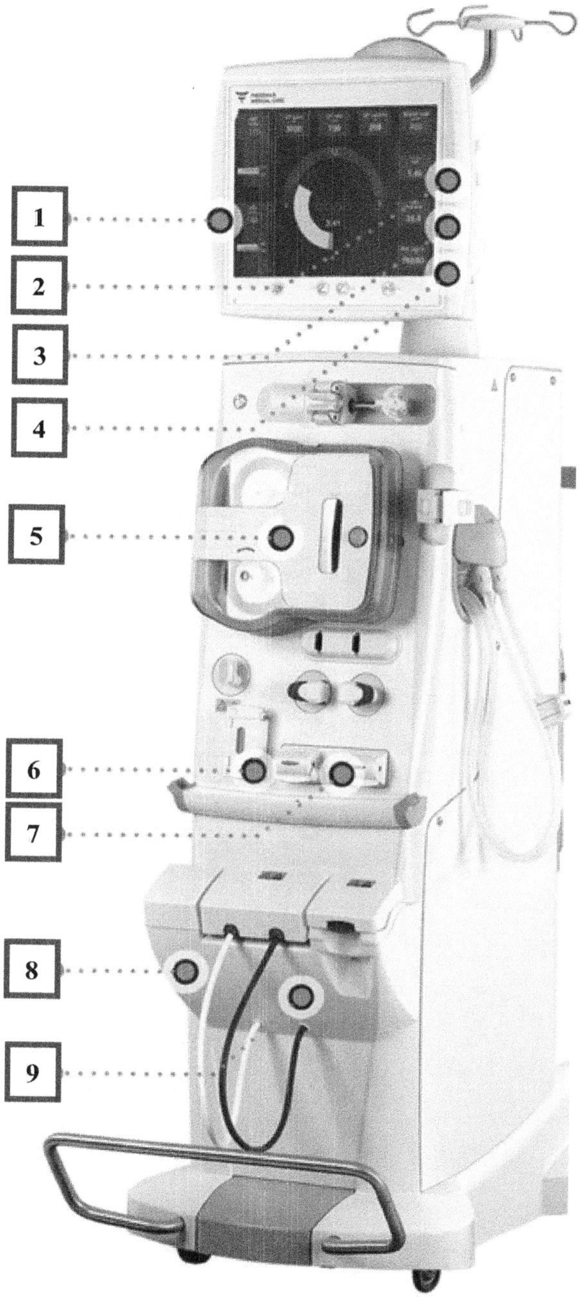

Figure 2.3. *Front view of the Fresinus hemodialysis machine*

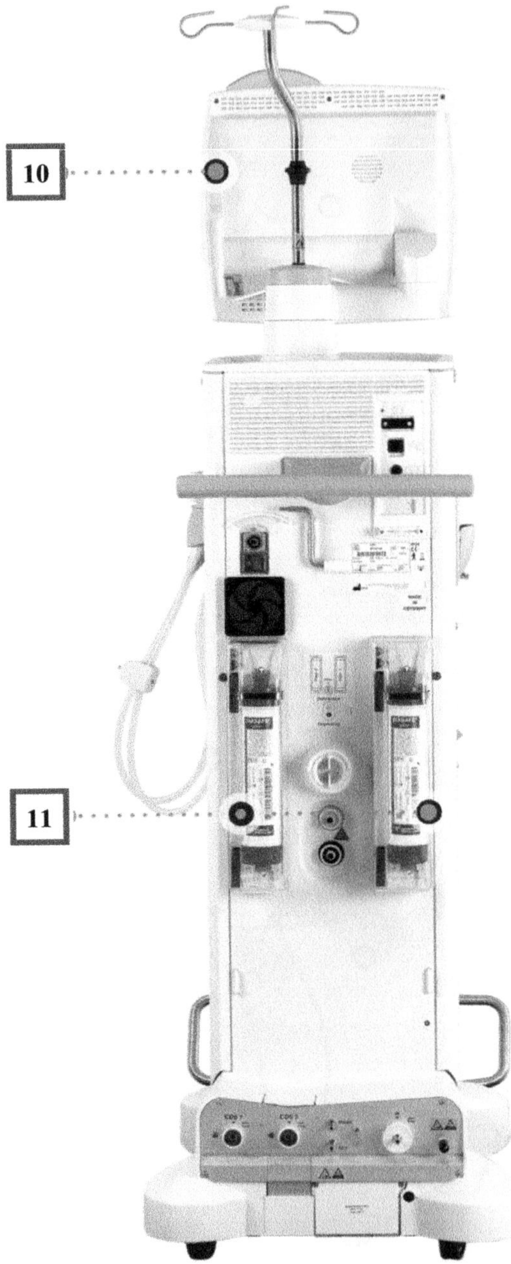

Figure 2.4. *Rear view of the Fresinus hemodialysis machine*

No.	Name	Role
1	**VAM (Venous Access Monitor)**	A safety system that continuously monitors venous pressure. It instantly detects a dislodged needle or a line break and clamps blood circulation to prevent significant blood loss.
2	**Emergency button**	A single button that, in an emergency, simultaneously triggers four critical actions: stops the blood pump, clamps lines, stops infusion, securing the patient and simplifies incident response.
3	**OCM (Online Clearance Monitor)**	Measures the effective dialysis dose (Kt/V) in real time via conductivity, eliminating the need for blood sampling. This means immediate treatment adjustments to ensure efficacy and individualization.
4	**AutoSub plus**	An intelligent algorithm that automatically calculates and applies the maximum and personalized replacement flow rate permitted by the transmembrane pressure limits, optimizing ultrafiltration efficiency and treatment time.
5	**Streamlined design**	An intuitive user interface with a touchscreen and logical workflow that reduces the number of clicks and steps required to configure and monitor a treatment. This minimizes errors and saves time.
6	**BVM (Blood Volume Monitor)**	Monitors, in real-time, changes in blood volume via optical sensors. It guides the clinician to optimize ultrafiltration management and prevent intradialytic hypotension.
7	**BTM (Blood Temperature Monitor)**	Constantly regulates blood temperature to maintain an isothermal state. This precise control improves the patient's cardiovascular tolerance and prevents overheating, contributing to comfort and hemodynamic stability.
8	**AutoFlow**	Automatically adjusts the dialysate flow according to the patient's actual needs to maintain optimal clearance. This function allows for substantial dialysate savings without treatment efficiency being compromised.
9	**EcoFlow**	An economical mode that intelligently reduces the machine's water and electricity consumption between treatments or during breaks, aligning clinical performance with environmental responsibility.
10	**IT workflow support**	A digital system that integrates with hospital IT to automate treatment data recording and analysis, streamlining retrospective documentation and improving accuracy.
11	**DIASAFE plus**	An advanced filter ensuring the produced dialysis fluid is sterile and non-pyrogenic, safeguarding patient safety by removing contaminants and endotoxins.

Table 2.1. *Nomenclature of the Fresinus hemodialysis machine*

3

Weibull Distribution

3.1. Introduction

In probability theory and statistics, the Weibull distribution is a continuous probability distribution widely applied in modeling lifetime data. Over the years, it has become one of the most cited distributions in reliability engineering and related disciplines, effectively capturing the failure times of diverse components and systems. For engineering analyses, it is essential to choose the distribution that best fits the observed dataset (O'Connor and Kleyner 2012). In industrial settings, the Weibull distribution ranks among the most frequently employed probability density functions. Lyonnet (1991) emphasized its particular suitability for evaluating the reliability of mechanical parts. A notable advantage of this distribution is its ability to work well, even with small datasets of failure observations. Nonetheless, when the number of observations is limited, graphical techniques are advised to improve the accuracy and precision of parameter estimation (Zhang and Xie 2007).

The Weibull distribution is highly valued for its flexibility in modeling different failure modes and for its ability to simulate many other statistical distributions (Lihou and Spence 1988). It is widely employed in life data analysis and can be applied to diverse practical situations. Applications of the Weibull distribution span multiple domains, including aerospace, electronics, materials, automotive industries and civil aviation (Luko 1999). Furthermore, this statistical tool provides a robust method for evaluating the reliability of components such as semiconductors, ball bearings, engines, spot welds and even biological systems (Rausand et al. 2020).

3.2. Description of the Weibull distribution

The Weibull distribution is characterized by three parameters (β, η and γ):

– β: this is the shape parameter (β > 0). This parameter gives indications on the failure mode and on the evolution of the failure rate over time.

- β < 1: λ (t) decreases over time (early life);
- β = 1: λ (t) is constant (useful life);
- β > 1: λ (t) is increasing (aging), in effect:
  - 1.5 < β < 2.5: expresses a phenomenon of fatigue;
  - 3 < β < 4: expresses a phenomenon of wear.

– η is the scale parameter.

– γ is the location parameter.

The failure rate is formulated through the two parameters of the Weibull distribution β and η (the location parameter γ = 0) by the following function:

$$\lambda(t) = \frac{\beta}{\eta}\left(\frac{t}{\eta}\right)^{\beta-1} \quad [3.1]$$

The reliability function can be given as follows:

$$R(t) = \exp\left(-\left(\frac{t}{\eta}\right)^{\beta}\right) \quad [3.2]$$

The cumulative density function is expressed in the following form:

$$F(t) = 1 - \exp\left(-\left(\frac{t}{\eta}\right)^{\beta}\right) = 1 - R(t) \quad [3.3]$$

The probability density function is written as:

$$f(t) = \frac{\beta}{\eta}\left(\frac{t}{\eta}\right)^{\beta-1} \exp\left(-\left(\frac{t}{\eta}\right)^{\beta}\right) = \lambda(t) \times R(t) \quad [3.4]$$

The failure of an equipment is characterized by the failure rate, which represents the frequency with which a system or component fails. If we represent the failure rate as a function of time, we obtain a curve called "bathtub curve". The curve in Figure 3.1 shows the evolution of the failure rate for the different systems. The failure rate curves have the same general shape, but nevertheless present differences according to the main technology of the studied system.

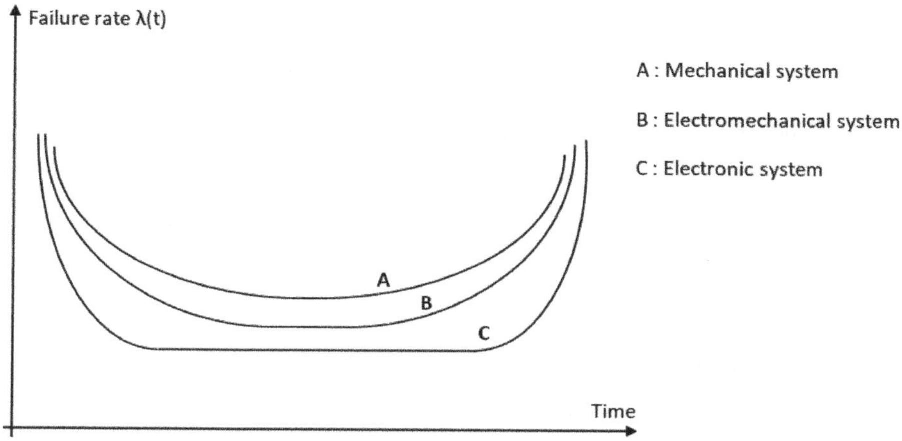

Figure 3.1. *Failure rate for different systems*

The figure of the bathtub curve below explains the different causes and solutions of failures linked to the different phases of the equipment's lifespan (Figure 3.2).

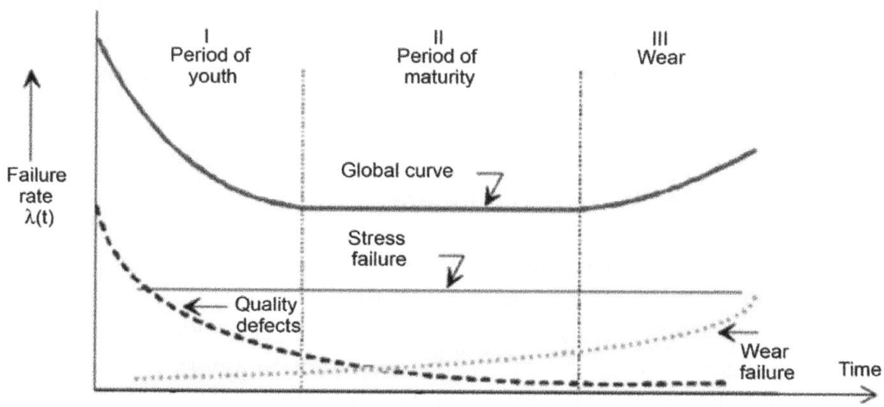

Figure 3.2. *The bathtub curve. For a color version of this figure, see www.iste.co.uk/bouchoucha/optimizing.zip*

The failure rate curve, a bathtub curve, can be constructed by associating three Weibull-type rates: decreasing ($\beta < 1$), constant ($\beta = 1$) and increasing ($\beta > 1$) (Ahsan et al. 2019). Figure 3.3 shows the evolution of the failure rate as a function of the shape parameter (β).

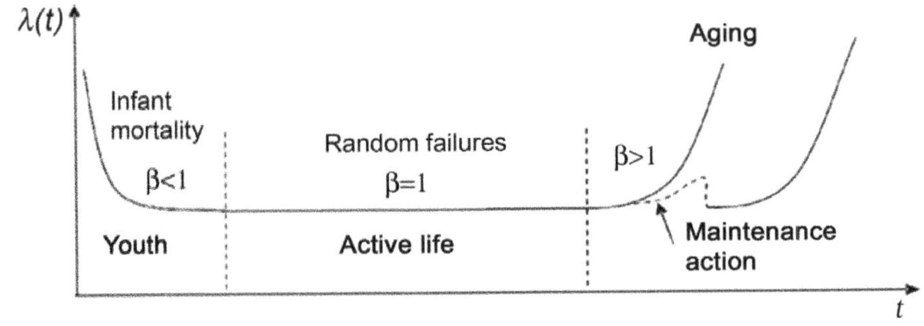

Figure 3.3. *Bathtub curve by the Weibull distribution*

3.3. Weibull distribution characteristics

3.3.1. *Advantages*

The Weibull distribution's main strength lies in its adaptability to different failure modes, making it a go-to model in engineering disciplines. It is suitable for small sample sizes and easy to estimate via graphical or numerical methods such as maximum likelihood estimation (MLE) (Le Son et al. 2013). As it can accurately model random failures, wear-out mechanisms and early-life failures, it finds use in lifecycle modeling across a range of industries.

3.3.2. *Limitations and disadvantages*

Despite its flexibility, the Weibull distribution assumes homogeneity of the failure mechanism, which can be unrealistic for systems with multiple degradation processes. When the hazard function is non-monotonic, such as in bathtub-shaped failure rates, the Weibull model underperforms unless extended into more complex forms, such as mixture Weibull models or Weibull–gamma distributions (Shahzad et al. 2019). Parameter estimation can also become unreliable when data are heavily censored or incomplete.

3.3.3. *Application domains*

The Weibull distribution is ubiquitous in mechanical and biomedical engineering, electrical systems and materials science. For instance:

– in *reliability-centered maintenance*, it is employed to model the lifetimes of bearings, valves and electronic components (Fidanoglu et al. 2017);

– in *biomedical engineering*, it supports analysis of survival times of pacemakers and joint implants;

– in *wind energy systems*, it is often employed to model wind speed distributions, aiding in the prediction of energy output (Luko 1999).

3.3.4. *Real-world example*

A notable application of the Weibull distribution is found in the failure analysis of turbine blades in jet engines. Maintenance records over a five-year period indicated that the failure times followed a Weibull distribution with $\beta > 1$, confirming a wear-out failure mode. Based on this, preventive replacements were scheduled before the mean time to failure, significantly reducing in-flight incidents and improving operational safety (Fidanoglu et al. 2017).

3.4. Estimation of the Weibull parameters

The Weibull distribution is highly flexible, capable of representing a wide variety of reliability patterns depending on its parameter values. It can be specified using either two or three parameters: scale, shape and location. The two-parameter version, in particular, is a continuous probability distribution widely employed for the analysis of reliability and lifetime data (Pasha et al. 2006).

Various analytical methods are available for estimating Weibull parameters, including the maximum likelihood method (MLM), the least squares method (LSM) and the method of moments (MOM) (Stone and Van Heeswijk 2007). These approaches are generally regarded as more precise and dependable compared to graphical methods (GM), which, although quicker, tend to be less accurate in parameter estimation (Lyonnet 1991).

3.4.1. *Least square method*

The objective of regression is to determine a relationship that may possibly govern two sets of data, $x = \{x_1, x_2, ..., x_n\}$ and $y = \{y_1, y_2, ..., y_n\}$, with size n, measured on a population or obtained experimentally. When the relationship being investigated is linear (y = ax + b), the problem is referred to as linear regression. However, even if such a relationship exists, the observed data rarely exactly conform to it. To account for these discrepancies in the mathematical model, the measured values of x and y are treated as realizations of two random variables, X and Y. The LSM is based on the principle of selecting the values of a and b, which minimize the sum of the squares of the deviations between the observed and predicted values.

$$E = \sum_{i=1}^{n} \varepsilon_i^2 = \sum_{i=1}^{n} \left(y_i - (ax_i + b)\right)^2 \quad [3.5]$$

where ε_i is the difference observed between y_i and $(ax_i + b)$, and

$$a = cov_{xy}/s_x^2 \quad [3.6]$$

$$b = \bar{y} - a\bar{x} \quad [3.7]$$

where cov_{xy} is the covariance and s_x^2 is the variance given by:

$$\text{cov}_{xy} = \frac{1}{n}\sum_{i=1}^{n}(x_i - \bar{x})(y_i - \bar{y}) \text{ and } s_x^2 = \frac{1}{n}\sum_{i=1}^{n}(x_i - \bar{x})^2 \quad [3.8]$$

where

$$\bar{x} = \frac{1}{n}\sum_{i=1}^{n} x_i \text{ and } \bar{y} = \frac{1}{n}\sum_{i=1}^{n} y_i \quad [3.9]$$

The parameters of the Weibull distribution can be estimated using the LSM. Specifically, the shape parameter β and the scale parameter η can be determined through the following mathematical procedure (Hadroug et al. 2016):

$$\beta = \sum_{i=1}^{n}(x_i - \bar{x})(y_i - \bar{y}) / \sum_{i=1}^{n}(x_i - \bar{x})^2 \quad [3.10]$$

$$\eta = \bar{y} - \beta\bar{x} \quad [3.11]$$

where

$$x_i = \text{Ln}(\text{TBF}_i), \qquad y_i = \text{Ln}(\text{Log}(1/1 - F(t_i))),$$

$$\bar{x} = \frac{1}{n}\sum_{i=1}^{n} \text{Ln}(\text{TBF}_i), \qquad \bar{y} = \frac{1}{n}\sum_{i=1}^{n} \text{Ln}(\text{Ln}(1/1 - F(t_i)))$$

Failure times, denoted as TBFi, were collected and sorted in ascending order to compute the cumulative distribution function (CDF), represented by $F(t_i)$. The Benard's formula (Pasha et al. 2006) was applied to assign median ranks to each observed failure, as presented in Table 3.1 (Cohen and Whitten 1982).

Method	Formula $F(t_i)$	Recommended for
Mean ranking	$F(t_i) = {}^{i}/_{n+1}$	$n > 50$
Median rank	$F(t_i) = {}^{i-0.3}/_{n+0.4}$	$n < 20$
Systematic	$F(t_i) = {}^{i-0.5}/_{n}$	$20 < n < 50$

Table 3.1. *Formulae available for median ranking*

Here, in the table, *n* is the total number of data and *i* is the order of failure.

Benard's formula provides a reliable approximation of the median rank estimator (Pasha et al. 2006). It demonstrates strong performance and is the most commonly used method for estimating F(ti) (Kim and Yum 2008). The procedure for ranking complete failure data involves first ordering the time-to-failure values from smallest to largest, and then applying Benard's formula to assign median ranks to each observed failure.

3.4.2. *Maximum likelihood method*

The MLM is an algorithm designed to fit observed data by determining the parameter values that maximize the likelihood function. For the two-parameter Weibull distribution, the shape parameter β and the scale parameter η can be estimated using this approach for the probability density function (Cohen and Whitten 1982) as follows:

$$f(t) = \left(\frac{\beta}{\eta}\right)\left(\frac{t}{\eta}\right)^{\beta-1} \exp(-\left(\frac{t}{\eta}\right)^{\beta}) \quad [3.12]$$

The maximum likelihood estimators for β and η are determined by maximizing the following function with respect to these parameters:

$$L(\beta, \eta) = \prod_{i=1}^{n} f(t_i, \beta, \eta) \quad [3.13]$$

The logarithmic likelihood is given in the following equation:

$$\mathrm{Ln}(L(\beta, \eta)) = \sum_{i=1}^{n} \mathrm{Ln}(f(t_i, \beta, \eta)) \quad [3.14]$$

To obtain the maximum likelihood estimators, the following system of equations needs to be solved:

$$\partial \mathrm{Ln}(L(\beta, \eta))/\partial \beta = 0 \quad [3.15]$$

$$\partial \mathrm{Ln}(L(\beta, \eta))/\partial \eta = 0 \quad [3.16]$$

Solving this system allows the derivation of the logarithmic likelihood function for the two-parameter Weibull distribution:

$$\mathrm{Ln}(L(\beta, \eta)) =$$

$$n\mathrm{Ln}(\beta) - n\mathrm{Ln}(\eta) - \sum_{i=1}^{n}(t_i/\beta)^{\eta} + (\beta - 1)\sum_{i=1}^{n} \mathrm{Ln}(t_i/\beta) \quad [3.17]$$

The following system can be extracted (Kappenman 1985; Ferreira and Silva 2017):

$$\beta = \left[\left(\sum_{i=1}^{n} t_i^{\beta} \mathrm{Ln}(t_i)\right)\left(\sum_{i=1}^{n} t_i^{\beta}\right)^{-1} - \frac{1}{n}\sum_{i=1}^{n} \mathrm{Ln}(t_i)\right]^{-1} \quad [3.18]$$

$$\eta = \left[\frac{1}{n}\sum_{i=1}^{n} t_i^{\beta}\right]^{\frac{1}{\beta}} \quad [3.19]$$

Solving this system using the Newton–Raphson method enables the calculation of the shape parameter β and the scale parameter η.

3.4.3. *Method of moments*

The MOM is widely employed due to its simplicity and ease of application. It also provides parameter estimates in situations where the MLM fails to converge. Additionally, the method of moments can serve as an initial step for setting parameter values in iterative procedures used by the maximum likelihood approach. The theoretical first- and second-order moments of the Weibull distribution are given, respectively, as follows (Annis 2006):

$$\mu_1(X) = \eta\Gamma(1 + 1/\beta) \quad [3.20]$$

$$\mu_2(X) = \eta^2[\Gamma(1 + 2/\beta) - \Gamma^2(1 + 1/\beta)] \quad [3.21]$$

where Γ is the gamma function given in Appendix 1.

When applied to the two-parameter Weibull distribution, the method of moments involves equating the sample's first and second moments (mean and variance) with the corresponding theoretical moments (μ_1 and μ_2).

$$\bar{x} = \mu_1(X), S^2 = \mu_2(X) \quad [3.22]$$

where:

$$\bar{x} = \frac{1}{n}\sum_{i=1}^{n} x_i \text{ and } s^2 = \frac{1}{n}\sum_{i=1}^{n}(x_i - \bar{x})^2$$

The following system of equations can be deduced from the previous equations:

$$\eta\Gamma(1 + 1/\beta) = \frac{1}{n}\sum_{i=1}^{n} x_i \quad [3.23]$$

$$\eta[\Gamma(1 + 2/\beta) - \Gamma^2(1 + 1/\beta)]^{\frac{1}{2}} = \left[\frac{1}{n}\sum_{i=1}^{n}(x_i - \bar{x})^2\right]^{\frac{1}{2}} \quad [3.24]$$

$$\left([\Gamma(1 + 2/\beta) - \Gamma^2(1 + 1/\beta)]^{\frac{1}{2}}/\Gamma(1 + 1/\beta)\right)\left(\frac{1}{n}\sum_{i=1}^{n} x_i\right) = \left[\frac{1}{n}\sum_{i=1}^{n}\left(x_i - \frac{1}{n}\sum_{i=1}^{n} x_i\right)^2\right]^{\frac{1}{2}} \quad [3.25]$$

This system cannot be solved directly and requires an iterative approach. The Newton–Raphson method can be employed to estimate β, after which the scale parameter η is calculated as a function of β.

3.4.4. *Interpretation and comparative analysis*

The estimation of Weibull distribution parameters – shape (β) and scale (η) – is a crucial step in reliability analysis and lifetime modeling. Among the methods discussed, each presents specific strengths and weaknesses that must be considered depending on the context and data characteristics.

The LSM is appreciated for its simplicity and clear analytical framework. It works well when the failure data can be linearized, particularly by applying transformations such as the double logarithm of cumulative distribution values. However, its reliance on graphical interpretation and sensitivity to data dispersion can reduce its accuracy in the presence of outliers or small sample sizes.

In contrast, the MLM is statistically optimal under the assumption that the underlying distribution is correct. It provides asymptotically unbiased, consistent and efficient estimators of the Weibull parameters. Its major advantage lies in its robustness and precision, especially with large datasets. However, its major limitation is computational complexity, as it requires solving a system of nonlinear equations, often by iterative algorithms such as Newton–Raphson algorithm.

The MOM offers a balance between simplicity and efficiency. It is particularly useful when MLM fails to converge or is too computationally intensive. MOM provides good initial estimates that can be used to initiate iterative optimization processes. Nevertheless, its estimators are generally less accurate than those from MLM, especially when data are sparse or heavily skewed, and it assumes the availability of sufficiently large and representative samples.

3.5. Conclusion

In conclusion, there is no universally best method for estimating the Weibull parameters; the choice depends largely on the data size, desired accuracy, computational resources and presence of censored data. For large datasets where precision is critical, MLM is the preferred method. For preliminary or quick analyses, or where visual diagnostics are useful, LSM offers a practical alternative. MOM serves as an efficient compromise or an initialization step for more advanced methods. A hybrid approach, combining these methods, is often beneficial – using MOM for initial values, MLM for refinement and LSM for graphical validation – especially in real-world reliability engineering contexts such as failure analysis in biomedical or mechanical systems.

4

Reliability, Maintainability and Availability Analysis Through Weibull Distribution

4.1. Introduction

Medical equipment maintenance can be defined as the set of activities required to keep medical devices in, or restore them to, a specified operational condition (Dhillon 2011). Numerous methods and procedures have been developed to enhance equipment reliability through mathematical maintenance models. For instance, Khalaf and Hamam (2013) proposed a model that evaluates the impact of maintenance on the survival probability of medical equipment by considering both maintenance history and equipment age. Similarly, Bahreini et al. (2018) investigated the factors influencing the management of medical equipment maintenance.

Reliability is generally described as the probability that a system will perform its intended function for a specific duration under defined operating conditions (Lyonnet 1991; Dhillon 2011). These conditions may include mechanical, chemical, atmospheric, human or other external constraints. Dynamic reliability extends the classical notion of reliability by analyzing failure behavior over time (Bertsche 2008). This approach accounts for parameter variations that affect system performance (such as dynamic processes or aging), structural modifications or environmental stresses (Devooght 2002).

A comprehensive review of inspection and maintenance practices for medical devices was provided by Jamshidi et al. (2014). In the case of dialysis machines, Alnazier et al. (2013) examined both mechanical and electrical failures to identify their causes, propose appropriate corrective measures and suggest preventive strategies. Furthermore, Tsarouhas (2012) applied the Weibull distribution to assess and improve the reliability and maintainability of hemodialysis machines.

4.2. General concept of failure

The failure of medical equipment can compromise the effectiveness of healthcare services, pose serious risks to patients and even cause environmental harm (Zamzam et al. 2021). Failure is defined as a partial or total loss of the essential properties of a component, leading to a significant reduction or complete loss of its operational capacity. Such failures may arise from design flaws, manufacturing defects, installation errors or inadequate maintenance practices (Arunraj and Maiti 2007; Brissaud et al. 2007). All production systems are inherently subject to wear and aging processes (Valdez-Flores and Feldman 1989). These physical phenomena contribute to failures that often increase operational costs and compromise safety.

In the literature, several authors have proposed different classifications of failures (Deloux et al. 2008; Rausand et al. 2020) and have examined how the aging of mechanical systems affects their reliability (Alfares 1999; Gorjian et al. 2010). The degradation of components, influenced by factors such as operating conditions, fatigue, vibrations and external stresses, necessitates time-dependent modeling of failure rates. Mechanical parts are particularly affected by complex and diverse degradation mechanisms, including cracking, creep, wear and fatigue fracture (Gorjian et al. 2010). These mechanisms are influenced by multiple parameters such as material properties, geometric characteristics and external stresses (Aven et al. 2006).

Different approaches have been used to classify failures. For instance, Rausand et al. (2020) proposed a classification based on cause, timing, detectability and severity, while Deloux et al. (2008) distinguished between failures according to their causes and their effects on system performance. Classifying failures by cause separates random failures from systematic ones, whereas categorization by impact differentiates intermittent failures from more extensive ones.

It has also been observed that failure rates reported in reliability databases can vary significantly (Aupied 1993). These discrepancies can be attributed to numerous factors, including equipment design characteristics, operating conditions, environmental influences, the accuracy of reported data and the increasing complexity of reliability evaluations, particularly for entire systems compared to individual mechanical components.

4.3. Systems degradation models

Degradation models are mathematical frameworks used to describe how a system deteriorates over time. They serve as valuable tools for representing the condition of a production system, defining maintenance strategies, evaluating system availability and determining optimal maintenance intervals (Gölbaşı and Demirel 2014).

The degradation of components, driven by factors such as operating conditions, fatigue, vibrations and other stresses, requires time-dependent modeling of failure rates. Several studies have highlighted the significant influence of aging on the reliability of mechanical systems (Alfares 1999). Mechanical elements are subject to diverse and often complex degradation mechanisms, including cracking, creep, wear and fatigue fracture (Habchi and Barthod 2016). These processes depend on various parameters, such as material properties, dimensional characteristics and external loading conditions (Tebbi et al. 2005).

According to Gorjian et al. (2010), degradation models related to aging effects can be grouped into two categories: physical models and empirical models. They further distinguish three main types: the shock model, the wear model and a hybrid model that combines both shock and wear processes. Additionally, Alfares (1999) classifies degradation models into discrete and continuous types, depending on how the deterioration is represented over time.

4.4. Factors influencing the reliability

Influence factors can be defined as internal or external elements that affect the reliability of a system. Their impact may be either positive, by reducing the occurrence of failures, or negative, by diminishing system reliability. The type of factors observed often depends on the nature of the system under study, whether human, electrical/electronic or mechanical. Such factors can include human or organizational influences, as well as intrinsic or extrinsic characteristics of the system itself.

Brissaud et al. (2007) proposed a classification of influence factors according to the life stages of the system, distinguishing between design factors, manufacturing factors, installation factors, usage-related factors and maintenance-related factors.

Additional dimensions such as human and organizational aspects, which usually exert a broader and more systemic impact, have been highlighted by several authors (Aven et al. 2006; Cui and Li 2019).

For mechanical systems such as valves, springs, seals and gears, models were developed by the Naval Surface Warfare Center (NSWC) (Chesley Naval Surface Warfare Center Logistics Engineering Technology Branch 1998). In this context, influence coefficients are calculated using physical relationships that depend on a variety of parameters, including temperature, pressure, dimensions and the physical properties of fluids and materials.

In contrast, for electrical and electronic components, failure rates, often assumed constant over time, are generally expressed analytically as functions of predefined parameters.

Cui and Li (2019) proposed a reliability model adapted to aeronautical maintenance data, showing that certain on-board system components may be considered reliable depending on the condition of related components. In situations where physical relationships linking influence factors to failure rates are unknown, statistical approaches are employed to estimate correction coefficients.

Finally, in order to illustrate the various factors and their interactions that lead to variations in reliability, Ouakki (2011) proposed the use of a cause–effect diagram (Ishikawa diagram), which provides a visual representation of the underlying mechanisms (Figure 4.1).

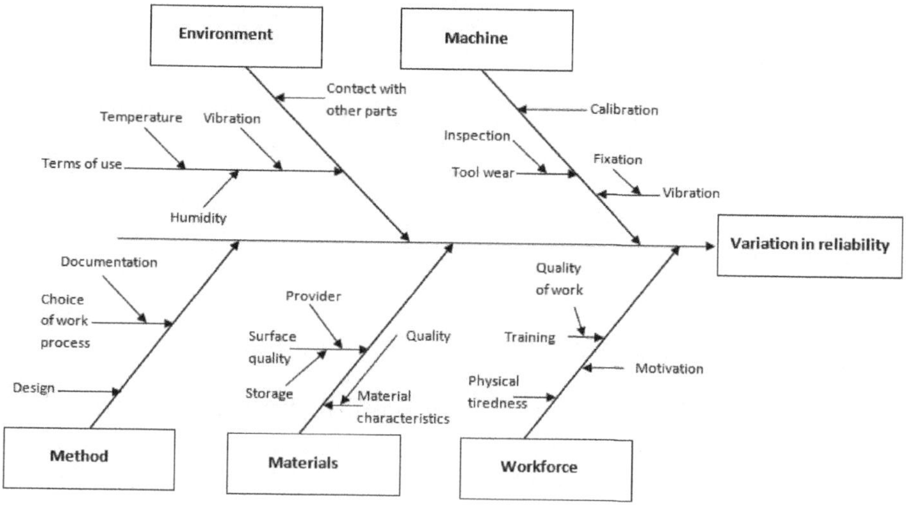

Figure 4.1. *Cause and effect diagram of the reliability*

4.5. Reliability, maintainability and availability analysis of medical equipment using the Weibull distribution

4.5.1. *Introduction*

The performance and safety of medical equipment play a crucial role in ensuring high-quality healthcare. Medical equipment maintenance encompasses all actions required to preserve or restore a device to its operational condition (Khalaf et al. 2010). In recent years, mathematical modeling has emerged as a vital tool in optimizing maintenance strategies. One of the most widely adopted statistical models for analyzing failure behaviors is the Weibull distribution, known for its flexibility in representing increasing, constant or decreasing failure rates depending on the system's aging process (Bahreini 2024).

Reliability is commonly defined as the probability that a system or component will accomplish its intended function without experiencing failure during a specified time interval under stated operating conditions (Wang and Pham 2006). Complementary to this concept, maintainability describes the ease and rapidity with which a system can be restored to its normal operational state after a failure occurs (Zio 2016). Availability, in turn, integrates both reliability and maintainability by expressing the proportion of time during which a system is in an operable condition and capable of fulfilling its required function (Xie et al. 2023).

These concepts are interconnected and critical in the biomedical domain, where equipment downtime can severely affect patient care. The Weibull distribution provides a robust analytical framework for quantifying reliability, optimizing maintenance scheduling and improving equipment availability (Saraswat and Yadava 2008).

4.5.2. *Application to reliability of medical equipment*

The operational demands placed on critical medical devices such as dialysis machines subject them to a combination of mechanical, electrical and environmental stressors that progressively degrade their components and performance. These include repeated thermal cycles from disinfection, chemical exposure to cleaning agents, vibration from pumps and valves, and electrical transients. Such conditions make reliability engineering not just beneficial but essential for maintaining continuous, safe patient care. Weibull analysis has emerged as a particularly powerful statistical tool in this context, as it effectively models time-to-failure data – especially for systems exhibiting clear wear-out phases – enabling a shift from reactive repairs to predictive maintenance strategies (Mutia et al. 2018).

By fitting historical and real-time failure data to the Weibull distribution, clinical engineering departments can move beyond scheduled time-based maintenance and accurately estimate the remaining useful life (RUL) of subsystems. This approach allows for the optimization of preventive maintenance intervals, ensuring interventions occur precisely when needed based on actual equipment usage and failure modes. The outcome is a significant reduction in unexpected breakdowns, which directly enhances treatment capacity, reduces operational costs and most importantly, mitigates risks to patient safety during dialysis sessions.

A practical case is the reliability monitoring of hemodialysis machine hydraulic pumps, which are critical for blood and dialysate flow. A Weibull shape parameter (β) value greater than 1, for instance, $\beta \approx 2.5$, strongly indicates a failure mode driven by mechanical wear, such as bearing degradation or seal fatigue. In such cases, performing maintenance or replacement just before the equipment reaches its characteristic life (η) – the point at which 63.2% of units in a population are expected to fail – can dramatically reduce the incidence of in-session failures. This proactive approach not only improves machine availability and reduces spare parts inventory through better planning but also strengthens overall healthcare delivery by ensuring that vital equipment is operational when needed (Mkalaf 2016). Thus, Weibull analysis serves as a cornerstone for advanced reliability-centered maintenance in modern clinical environments.

4.5.3. *Maintainability modeling using Weibull parameters*

Maintainability is measured by the mean time to repair (MTTR). While MTTR is often assumed to follow an exponential or log-normal distribution, recent works highlight the relevance of Weibull modeling for repair times in complex biomedical systems, where the duration of maintenance depends on the failure type and repair process maturity (Liao et al. 2021).

The maintainability function *M*(*t*), representing the probability that a repair is completed within time *t*, can also adopt a Weibull form:

$$M(t) = 1 - e^{-\left(\frac{t}{\eta}\right)^{\beta}} \qquad [4.1]$$

where η and β are repair-related scale and shape parameters. For devices with significant variance in repair duration, this model captures tail behavior more accurately than exponential models.

The application of the Weibull model for maintainability enables clinical engineers and equipment fleet managers to move beyond traditional approaches based on often imprecise average MTTR values. This refined modeling is particularly relevant for complex biomedical systems such as dialysis generators, where intervention times vary significantly depending on the nature of the failure.

Concretely, estimating η and β parameters specific to each equipment type and failure mode allows for:

– *realistic planning of technical resources* by identifying potentially time-consuming repairs (tail of the distribution);

– *optimization of technician training* by targeting interventions with high β values requiring specific expertise;

– *improvement of overall fleet availability* by reducing uncertainty regarding downtime durations

For critical equipment, this approach also enables the definition of maintenance contracts tailored to actual risks and the optimization of strategic spare parts inventory levels. Weibull modeling thus provides a robust decision-making framework for managing the lifecycle of medical devices, proving particularly valuable in a context of increasing budgetary constraints.

4.5.4. *Availability analysis and system performance*

Availability A is a function of both reliability and maintainability, often expressed as:

$$A = \frac{\text{MTBF}}{\text{MTBF+MTTR}} \quad [4.2]$$

where MTBF (mean time between failures) can be estimated using the Weibull reliability function:

$$\text{MTBF} = \int_0^\infty tf(t)dt = \eta\Gamma(1 + \frac{1}{\beta}) \quad [4.3]$$

with Γ being the Gamma function (Appendix 1).

Incorporating Weibull-based MTBF and MTTR into availability calculations allows healthcare facilities to benchmark equipment performance and prioritize investments in more reliable and maintainable technologies (Zamzam et al. 2021).

This mathematical framework moves beyond theoretical reliability engineering to provide actionable insights for clinical technology management. In healthcare settings, availability is the ultimate metric, directly impacting patient care capacity and operational efficiency. The presented formula, A = MTBF/(MTBF + MTTR), succinctly captures the tension between a device's inherent reliability (MTBF) and the efficiency of the support system in repairing it (MTTR).

The critical advancement here is the use of the Weibull distribution to calculate a more realistic MTBF. Unlike a simple arithmetic mean, the

Weibull-based MTBF ($\eta\Gamma(1+1/\beta)$) accounts for the device's specific failure behavior. For instance, a shape parameter $\beta > 1$ indicates wear-out, meaning the MTBF changes over time; this formula provides a weighted average that reflects this reality.

By integrating Weibull-derived values for both MTBF and MTTR, clinical engineers can perform nuanced analyses. For example:

– *Scenario modeling*: they can accurately predict how improving maintainability (e.g. by training technicians on specific high-β repairs) would impact overall availability, even without changing the device's inherent reliability.

– *Technology assessment*: when evaluating new dialysis machines, a facility can benchmark competing models not just on purchase price, but on their predicted lifecycle availability, factoring in the expected failure patterns and repair times unique to each technology.

This approach allows healthcare facilities to shift from reactive maintenance to strategic, data-driven asset management, prioritizing investments that truly maximize uptime and protect clinical workflow.

4.5.5. *Degradation modeling and preventive maintenance*

Degradation models describe how a system deteriorates over time. For medical equipment, degradation can be due to chemical corrosion, thermal stress or mechanical fatigue. Weibull-based degradation modeling supports condition-based maintenance (CBM), where maintenance actions are triggered by equipment condition rather than fixed schedules (Cao et al. 2025).

Advanced statistical techniques such as proportional hazards modeling (PHM) and Weibull regression further refine predictions by integrating covariates such as usage intensity, operator variability and environmental conditions (Hu and Chen 2020).

This approach marks a significant evolution from traditional preventive maintenance, which often relies on conservative, fixed-time intervals that can lead to either unnecessary maintenance or unexpected failures. Weibull-based degradation modeling provides a mathematical foundation for predicting the precise trajectory of a component's wear, moving from a schedule-based to a condition-based paradigm.

For instance, in a dialysis machine, the peristaltic pump is subject to mechanical degradation. By modeling the gradual increase in motor current required to maintain a constant blood flow rate – a key performance indicator – against a Weibull failure model, the system can predict the RUL of the pump assembly. This allows technicians to order parts and schedule replacement during a planned downtime, utterly avoiding a disruptive mid-treatment failure.

Furthermore, the integration of covariates through techniques like PHM is what transforms a general model into a precise, unit-specific prognostic tool. A machine used for 20 treatments per day in a humid environment will have a different degradation profile than one used for 10 treatments in a controlled climate. By incorporating these operational and environmental "covariates", the model personalizes predictions, ensuring that maintenance is performed exactly when and where it is needed most, optimizing resource allocation and maximizing equipment availability and patient safety.

4.5.6. *Case studies and practical implications*

A study on infusion pumps revealed a Weibull shape parameter $\beta = 1.8$, indicating that wear-out mechanisms dominate failure behavior. Implementing a Weibull-driven preventive maintenance policy reduced equipment failure rate by 30% and MTTR by 15% (Silva et al. 2025).

Similarly, a reliability-centered maintenance (RCM) program applied to dialysis equipment using Weibull analysis led to (Shamayleh et al. 2020):

– an increase in equipment availability from 92%–97%;

– a reduction in emergency interventions;

– extended mean operational life of critical components.

4.5.7. *Conclusion*

The application of the Weibull distribution to the analysis of reliability, maintainability and availability of medical equipment offers a rigorous and flexible approach to maintenance optimization. By enabling failure prediction, maintenance planning and risk mitigation, Weibull analysis supports decision-making for biomedical engineers, maintenance managers and hospital administrators.

Future research may extend this approach by integrating machine learning for parameter estimation, real-time degradation monitoring through IoT sensors and hybrid reliability models that combine physical and statistical modeling.

4.6. Failure mode and effects analysis, fault tree analysis and Bayesian networks for dialysis machine reliability

4.6.1. *Introduction*

Ensuring the reliability, safety and availability of dialysis machines is paramount due to their critical role in sustaining patients with renal failure. To achieve this, healthcare engineers employ a variety of risk analysis tools designed to anticipate failures, understand their consequences and mitigate their impacts. Among the most widely used tools are failure mode and effects analysis (FMEA), fault tree analysis (FTA), and Bayesian networks (BNs). Each method offers unique strengths and limitations, making them suitable for different stages and aspects of risk management in complex systems like dialysis machines. This chapter presents a comprehensive and comparative overview of these techniques, focusing on their principles, implementation, advantages, limitations and their specific applications to dialysis systems.

4.6.2. *Failure mode and effects analysis*

4.6.2.1. *Principle and Construction*

FMEA is an inductive, bottom-up approach that systematically evaluates potential failure modes within a system and their resulting effects. The method involves listing all components, identifying possible failure modes for each and analyzing the impact of each failure based on three criteria:

– *Severity* (*S*): the seriousness of the effect on the system or patient.

– *Occurrence* (O): the frequency with which the failure is expected to occur.

– *Detection* (D): the likelihood that the failure will be detected before causing harm.

The product of these criteria yields the risk priority number (RPN):

$$RPN = S \times O \times D \tag{4.4}$$

The true value of the RPN lies not in the absolute number itself, but in the structured, cross-functional discussion it necessitates and the comparative prioritization it enables. By forcing a multidisciplinary team – including clinical engineers, technicians and sometimes even clinicians – to numerically score each failure mode, FMEA transforms subjective concerns into objective, actionable data.

This process often reveals critical vulnerabilities that are not immediately obvious. For instance, a failure mode with moderate severity but very high occurrence and very low detection can yield a high RPN, highlighting a previously overlooked operational weakness that could lead to frequent treatment interruptions. Conversely, a high-severity event with extremely low occurrence and multiple detection layers might be deemed an acceptable risk.

The primary outcome is a prioritized action plan. Teams focus on mitigating high-RPN items by:

– *Reducing occurrence*: redesigning a component or improving manufacturing quality.

– *Improving detection*: adding new sensors, implementing software alarms or introducing new diagnostic tests during preventive maintenance.

– *Mitigating severity*: incorporating redundant safety systems that contain the failure should it occur.

Furthermore, an FMEA is a living document. It must be periodically revisited and updated with real-world failure data from the field, allowing the model to evolve from a predictive tool into a continuously improving accuracy engine for reliability management. This iterative process is fundamental for upholding the highest levels of patient safety and operational efficiency in clinical environments.

FMEA is a proactive, systematic and excellent tool for initiating risk analysis. Its main strengths are its simplicity and its ability to prioritize efforts. However, its subjectivity and static, isolated nature make it an incomplete tool. It should be combined with other methods (such as FTA for interdependencies) and regularly updated with field data to form a robust and sustainable risk management framework, especially for critical systems like dialysis machines.

4.6.2.2. *Advantages*

The advantages are as follows:

– *Ease of implementation and interpretation*: FMEA does not require complex mathematical models. Its tabular structure (component, failure mode, effect, cause, S, O, D, RPN) is intuitive and easily understood by multidisciplinary teams, including engineers, technicians and clinical staff. This accessibility makes it a foundational tool for initiating a culture of proactive risk management.

– *Usefulness in early design phases and preventive maintenance planning*: when applied early in equipment design (often called design FMEA), it allows teams to identify and correct potential flaws "on paper" before they become costly to fix during production. For maintenance, it is used to develop targeted preventive maintenance plans focused on the most critical components, thereby optimizing resources and preventing failures.

– *Prioritization of corrective actions*: calculation of the RPN (= Severity × Occurrence × Detection) provides a quantitative method (albeit based on subjective ratings) to rank risks. This enables teams to focus objectively on the most dangerous and/or most likely failures, rather than reacting randomly or based on intuition.

4.6.2.3. *Limitations*

Limitations are as follows:

– *Subjectivity in scoring*: the ratings assigned for severity (S), occurrence (O) and detection (D) are based on the judgment and experience of the team. Two different teams may assign significantly different scores for the same failure mode, leading to varying RPNs and thus different action priorities. This subjectivity can undermine the robustness and reproducibility of the analysis.

– *Fails to consider interdependencies*: FMEA examines failures in isolation, component by component. It does not capture cascade failures or complex interactions between failures of multiple components. For example, it cannot model how the failure of a sensor (X) could cause a pump (Y) to malfunction. Other methods, such as FTA, are required to analyze these complex scenarios.

– *Static analysis*: once completed, an FMEA is a "snapshot" of risk based on the system's knowledge at a given time. It is not dynamic; it does not automatically update with feedback from real-world data (e.g. occurred failures and new detection measures). To remain relevant, it must be manually and periodically reviewed, which is often time-consuming and tedious, making it unsuitable for real-time risk management.

4.6.2.4. *Applications in dialysis*

Component	**Failure mode**	**Effect**	**Severity (S)**	**Occurrence (O)**	**Detection (D)**	**RPN = S × O × D**
Blood pump	Inaccurate flow rate	Inadequate blood purification	8	5	4	160
Dialyzer membrane	Leakage or rupture	Blood contamination, infection risk	10	2	5	100
Ultrafiltration pump	Over-removal of fluid	Hypotension or dehydration	9	3	3	81
Conductivity sensor	Incorrect conductivity reading	Electrolyte imbalance in patient	7	4	4	112
Temperature sensor	Faulty temperature control	Patient burns or hypothermia	9	2	3	54
Alarm system	Failure to activate	No warning during abnormal conditions	10	1	7	70
Blood leak detector	Failure to detect blood leakage	Blood loss, risk of contamination	9	2	6	108
Power supply	Unexpected shutdown	Treatment interruption	8	2	5	80

Table 4.1. *Machine components, failure modes and corresponding RPN values*

FMEA has been used to assess risks in components like blood leak detectors, ultrafiltration pumps and venous pressure sensors. For example, an FMEA can help schedule frequent calibration of the blood leak detector due to its criticality in preventing patient blood loss (Table 4.1).

4.6.3. *Fault tree analysis*

4.6.3.1. *Principle and construction*

FTA is a top-down, deductive method used to analyze the causes of a specific undesirable event. The analysis begins with a top-level fault (e.g. "dialysis not delivered") and decomposes it into intermediate and basic events through logic gates such as AND and OR. These gates represent how combinations of events can lead to top-level failure.

This structured approach is invaluable for moving beyond simple failure lists to understanding the complex, interconnected causes of system-level faults. In the context of a dialysis machine, the top event "dialysis not delivered" is rarely caused by a single failure. FTA reveals how a combination of mechanical, electrical and human factors can converge to cause a critical shutdown.

For example, an OR gate beneath the top event might show that dialysis failure can occur due to either a "Power Loss" OR a "Blood Pump Failure". The "Blood Pump Failure" itself could be decomposed by an AND gate, requiring both a "Pump Motor Fault" AND a "Backup Battery Failure" to occur simultaneously for the pump to completely stop. This logic powerfully illustrates how redundancy (like a backup battery) can be designed to prevent a single point of failure from causing the top event.

The primary output of an FTA is a visual map of failure pathways, which allows engineers to:

– *Identify critical single points of failure*: it includes basic events that alone can cause the top event (via OR gates).

– *Quantify risk*: by assigning failure probabilities to basic events, the probability of the top event can be calculated.

– *Prioritize mitigation efforts*: focus resources on strengthening the most likely or most critical failure paths, thereby designing more robust and safer systems. This method is foundational for meeting stringent medical device safety standards.

FTA is an indispensable, rigorous tool for analyzing high-risk, low-probability events and is a cornerstone of safety engineering. Its primary strengths are its visual clarity, effectiveness in finding root causes and ability to model complex failure interactions quantitatively.

However, its application is best suited for focused, critical top events rather than entire systems due to its complexity, static nature and dependence on analyst completeness. For a comprehensive risk management strategy, FTA should be used in conjunction with other tools like FMEA (which is better for bottom-up, component-level analysis) to provide a complete picture of system reliability and safety.

4.6.3.2. *Advantages*

The following are the advantages:

– *Powerful visual representation of failure logic*: the fault tree diagram is an intuitive and logical map that visually traces the chain of events leading to a system failure. This graphical format is exceptionally effective for communicating complex failure interactions to stakeholders, including engineers, management and regulatory bodies. It transforms abstract failure concepts into a clear, understandable structure.

– *Exceptional for root cause analysis*: FTA's deductive, top-down nature is specifically designed to drill down to the fundamental causes (root causes) of a specified critical failure. By systematically asking "how can this happen?" for each event, it ensures that analysis moves beyond symptoms to identify the underlying basic faults in components, software, procedures or human actions.

– *Enables both qualitative and quantitative assessment*: FTA is a versatile tool. Qualitatively, the structure of the tree itself (the arrangement of AND/OR gates) immediately reveals critical single points of failure and common cause failures. Quantitatively, by assigning failure probabilities to the basic events, engineers can calculate the numerical probability of the top event occurring. This allows for data-driven risk assessment and cost–benefit analysis of potential design changes or safety measures.

– *Identifies critical paths and dependencies*: unlike methods that look at failures in isolation (like FMEA), FTA excels at showing how failures combine. It clearly identifies which combinations of events (via AND gates) are necessary for a failure and which single points of failure (via OR gates) are most critical. This is essential for designing effective redundancies and safety interlocks.

4.6.3.3. *Limitations*

The following are the limitations:

– *Complexity and scalability for large systems*: for complex systems like a full dialysis machine, the fault tree can become enormous, intricate and very difficult to manage manually. The number of branches and logic gates can explode, making the tree time consuming to build, prone to errors and challenging to interpret. Specialized software is often required, adding to the resource burden.

– *Static model lacking dynamic behavior*: a traditional FTA is a static snapshot of the system's failure logic at a single point in time. It does not easily account for:

- *Temporal sequences*: the order in which failures happen can be crucial, but FTA treats simultaneous occurrences.

- *Repairs and maintenance*: it assumes components are not repaired during the mission time.

- *Feedback loops and dynamic system responses*.

- *Uncertainties and variations* in failure data. This makes it less suitable for analyzing systems with time-dependent behaviors.

– *Requires complete and accurate fault path identification*: the accuracy and usefulness of an FTA are entirely dependent on the analyst's ability to identify every possible fault path and basic event. If a potential cause or interaction is overlooked during the tree's construction, it will not appear in the analysis, creating a false sense of security. This makes the process highly reliant on the expertise and experience of the team.

– *Can be time-consuming and resource-intensive*: constructing a thorough and accurate fault tree is not a quick process. It requires significant time from highly skilled personnel to brainstorm all the potential causes, define the logic gates correctly and gather reliable data for quantitative analysis. For this reason, it is typically reserved for analyzing critical, high-consequence failures rather than every possible system fault.

4.6.3.4. *Applications in dialysis*

FTA is effective in mapping causes for failures like "Venous return blockage", which may stem from simultaneous events like "Clot formation" AND "Sensor undetected". It aids in implementing targeted solutions such as better anticoagulation protocols and sensor redundancy (Figure 4.2).

Explanation of branches are as follows:

– *Power failure*:

- *UPS failure*: uninterruptible power supply fails during operation.
- *Cable loose*: power cord is disconnected or damaged.

– *Blood circuit failure*:

- *Pump failure*: blood pump stops or malfunctions.
- *Dialyzer leak*: membrane rupture or leak leading to contamination.

– *Monitoring failure*:

- *Sensors faulty*: conductivity or temperature sensors give incorrect data.
- *Alarms fail*: audio/visual warnings do not activate.

– *Fluid management failure*:

- *UF pump failure*: ultrafiltration pump removes too much or too little fluid.
- *Valves blocked*: flow-regulating valves fail or are obstructed.

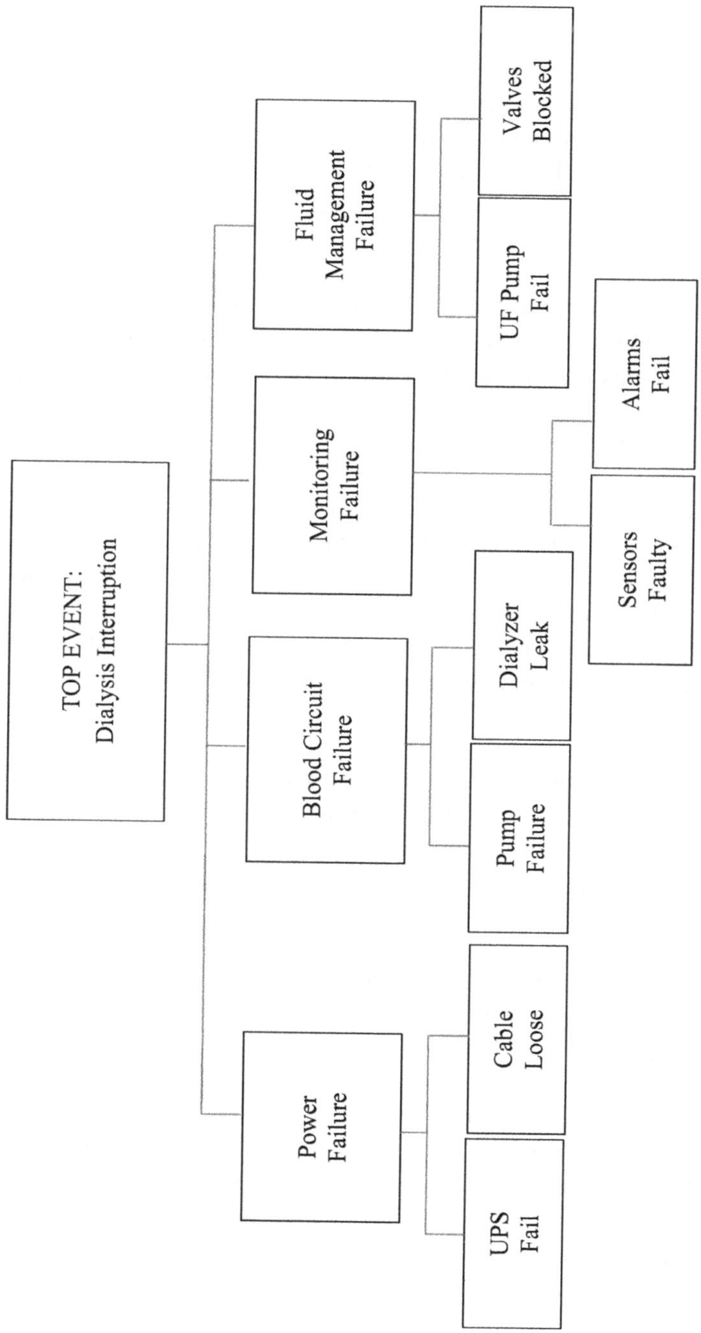

Figure 4.2. *A fault tree showing the top-level failure "Dialysis Interruption" decomposed into component-level faults*

4.6.4. *Bayesian networks*

4.6.4.1. *Principle and construction*

FTA and BNs are two powerful tools for modeling system reliability. The former relies on deterministic Boolean logic, structuring the causes of an undesired event using logic gates (AND and OR). It provides a clear representation of failure sequences, but assumes fixed relationships without explicitly incorporating uncertainty.

The BN, by contrast, models probabilistic relationships between events, integrating uncertainty and partial dependence. It enables dynamic updating of event probabilities based on observations. However, its implementation is more complex, requiring the definition of conditional probability tables for each dependent node.

BNs are probabilistic graphical models that capture a set of variables and their conditional dependencies using a directed acyclic graph (DAG). In these networks, each node corresponds to a random variable, such as a component's state or a sensor measurement, while the edges represent causal or dependency relationships. BNs enable reasoning under uncertainty by applying Bayes' theorem, which allows the probability of a given hypothesis to be updated as new evidence is incorporated.

BNs represent a significant leap forward in reliability modeling due to their ability to handle uncertainty, learn from new data and perform dynamic, real-time reasoning. They are arguably the most powerful tool for diagnosis, prognosis and real-time risk monitoring of complex systems like medical devices.

However, this power comes at a high cost: extensive development effort, deep expertise and potential computational demands. They are not a "quick and easy" solution.

Therefore, BNs are best deployed for modeling critical, high-value subsystems where their ability to predict and diagnose faults justifies the initial investment. They are often used to augment and add dynamic capabilities to other models, such as converting a static fault tree into a BN for more advanced analysis.

4.6.4.2. *Advantages*

The following are the advantages:

– *Handles incomplete and uncertain data*: unlike deterministic models that require precise inputs, BNs are designed to work with probabilistic data. They can incorporate uncertain information (e.g. "there's an 80% chance the sensor is faulty") and still provide useful output probabilities. This is invaluable in real-world engineering where data are often imperfect or scarce.

– *Allows real-time and dynamic risk assessment*: this is a key differentiator from static methods like FTA. A BN can act as a "living model". As new evidence is observed (e.g. a sensor reading changes, a maintenance action is logged), the probabilities of all connected nodes in the network are updated in real-time using Bayes' theorem. This enables predictive maintenance by forecasting failures before they occur and allows for dynamic reassessment of risk during operation.

– *Integrates expert knowledge with empirical data*: BNs provide a formal framework to combine two crucial types of information. Where historical data are lacking, you can populate the conditional probability tables (CPTs) with estimates from domain experts. As operational data are collected, the model can be refined and "learn" to become more accurate over time, balancing expert judgment with real-world evidence.

– Enables probabilistic inference for diagnosis and prognosis: BNs can reason in both directions:

– *Diagnosis (bottom-up)*: given observed symptoms (e.g. "low dialysate flow"), the network can calculate the most probable root cause (e.g. "clogged filter" vs. "pump failure").

– *Prognosis (top-down)*: given a known state of a component (e.g. "pump bearing worn"), the network can predict the probability of future system failures (e.g. "complete pump seizure"). This makes them powerful for both troubleshooting and forecasting.

– *Models complex, interdependent relationships*: BNs can represent dependencies that are impossible or overly complex to model in a fault tree. For example, they can easily handle common cause failures (a single event affecting multiple components) and circular or feedback relationships, providing a more realistic model of complex system interactions.

4.6.4.3. *Limitations*

The following are the limitations:

– *Requires complex modeling and significant expert input*: building an accurate BN is non-trivial. The model's structure (the DAG) must correctly represent all causal relationships. Furthermore, defining the conditional probability tables (CPTs) for each node is a massive effort. For nodes with multiple parents, the CPT can become enormous, requiring a vast number of probability assessments, often from busy and expensive domain experts.

– *Data collection can be challenging*: while BNs can handle some uncertainty, their accuracy is ultimately dependent on the quality of the data used to define the CPTs. For new systems or rare failure modes, obtaining sufficient reliable data to quantify these probabilities is a significant challenge. "Garbage in, garbage out" very much applies.

– *Computationally intensive for large systems*: probabilistic inference – updating the probabilities across the entire network when new evidence is entered – can require intense computation, especially for large, densely connected networks. While algorithms exist, they can be slow for real-time applications on very large models, potentially requiring specialized hardware or simplified models.

– *Requires deeper knowledge of probability theory*: effectively constructing, interpreting and validating a BN requires a stronger foundational understanding of probability theory and Bayesian statistics than simpler methods like FMEA or even FTA. This can create a barrier to adoption for teams without this specific expertise.

– *Complexity of construction for large systems*: as the system being modeled grows, the BN can become incredibly complex and difficult to build, manage and validate. The model can become a "hairball" of connections that is hard to understand, making it prone to errors in its structure that can invalidate the results.

4.6.4.4. *Applications in dialysis*

4.6.4.4.1. Transforming fault tree into a BN

The top-level event is as follows (Figure 4.3):

– Dialysis interruption

This depends on three direct causes connected via an OR gate:

– Power failure.

– Pump alarm.

– Line blockage.

Each of these causes is further segmented:

– Power failure: is caused by battery failure OR power supply fault (OR gate).

– Pump alarm: is caused by pump motor stuck OR pump controller fault (OR gate).

– Line blockage: is caused by line clamp failure OR blood leak detector fault (OR gate).

4.6.4.4.2. Building the BN

Steps are as follows:

– Each event (basic or intermediate) becomes a node.

– OR gates are translated into arcs pointing to the same node, indicating multi-source dependence.

– The tree is transformed into a DAG, where arcs point toward the event they influence.

4.6.4.4.3. Network structure

Level 1 (root causes) includes the following:

– Battery failure.

– Power supply fault.

– Pump motor stuck.

– Pump controller fault.

– Line clamp failure.

– Blood leak detector fault.

Level 2 (intermediate events) includes the following:

– Power failure ← [Battery failure, power supply fault].

– Pump alarm ← [Pump motor stuck, pump controller fault].

– Line blockage ← [Line clamp failure, blood leak detector fault].

Level 3 (top event) includes the following:

– Dialysis interruption ← [Power failure, pump alarm, line blockage].

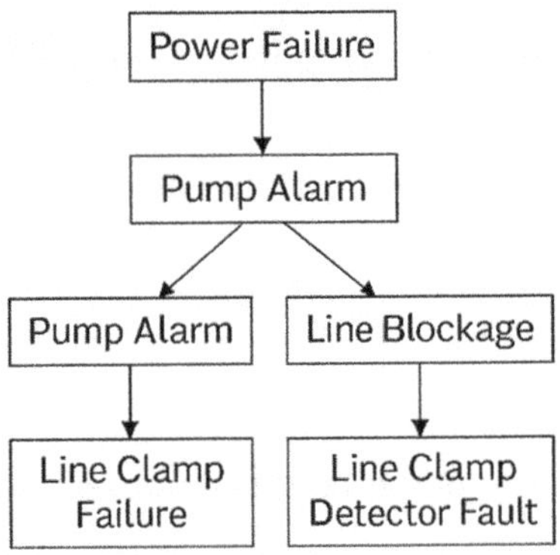

Figure 4.3. *Bayesian network showing the top-level failure "Dialysis Interruption" decomposed into component-level faults*

4.6.5. Comparative summary and relevance to dialysis machines

Criterion	FMEA	FTA	Bayesian networks
Approach	Bottom-up (inductive)	Top-down (deductive)	Probabilistic and dynamic
Dependency modeling	No	Limited (logical)	Yes (conditional probability)
Real-time updates	No	No	Yes
Ease of use	High	Moderate	Low (requires expertise)
Application stage	Design, maintenance	Root cause analysis	Predictive diagnostics

Table 4.2. *Comparative summary of the reliability tools: FMEA, FTA and BN*

Using a combination of these methods allows for a robust and holistic risk assessment strategy. FMEA helps prioritize initial hazards, FTA maps out logical sequences of failures and BNs provide adaptive responses based on real-time data (Table 4.2).

4.6.6. *Conclusion*

The comprehensive analysis presented in this chapter demonstrates that FMEA, FTA and BNs each offer distinct and complementary approaches for enhancing the reliability of dialysis machines. FMEA provides a systematic foundation for initial risk identification and prioritization through its bottom-up, component-level analysis. FTA contributes a deductive, top-down methodology for understanding complex fault interactions and root causes through logical structuring of failure pathways. Most importantly, BNs enable dynamic, probabilistic reasoning capable of integrating real-time data, managing uncertainties and supporting both diagnostic and prognostic decision-making.

The true power of these methodologies emerges not from their isolated application, but from their strategic integration across the equipment lifecycle. FMEA guides initial design improvements and maintenance planning, FTA reveals critical system vulnerabilities and common-cause failures, while BNs provide operational intelligence for predictive maintenance and real-time risk assessment. Together, they form a multi-layered framework that transforms reliability management from reactive troubleshooting to proactive prediction and prevention.

For healthcare technology managers, this integrated approach enables data-driven decision-making that enhances both patient safety and operational efficiency. By implementing these methods synergistically, medical facilities can significantly improve equipment availability, optimize maintenance resources, reduce lifecycle costs and ultimately ensure the highest standards of care for dialysis patients. This holistic reliability strategy represents not merely a technical improvement but also a fundamental advancement in healthcare technology management that directly contributes to better patient outcomes and more sustainable healthcare delivery systems.

4.7. Probabilistic analysis of dialysis interruption: fault tree and Bayesian network approaches with Weibull distribution integration

4.7.1. *Introduction: Weibull distribution and biomedical equipment*

Biomedical systems such as dialysis machines consist of critical components whose failure can result in immediate treatment interruption, posing significant risks to patient safety. The Weibull distribution is widely used to model the reliability behavior of such components:

$$\mathrm{R}_i(t) = \exp\left(-\frac{t}{\eta_i}\right)^{\beta_i} \text{ and } F_i(t) = 1 - R_i(t) \quad [4.5]$$

where:

– $R_i(t)$ is the reliability function of component i;

– $F_i(t)$ is the cumulative failure function;

– η_i is the scale parameter (characteristic life);

– β_i is the shape parameter (indicating infant mortality, random failure or wear-out).

This distribution allows the modeling of various aging behaviors, making it suitable for representing the operational realities of biomedical components (Hsu et al. 2024). These reliability functions can be embedded into fault tree and BN models. Their application is illustrated below through the analysis of a dialysis interruption scenario (Figure 4.2).

4.7.2. *Case study: dialysis interruption*

In Figure 4.2, the top event is a *dialysis interruption*, resulting from a chain of component failures:

– circulation pump (P);

– leak detector (L);

– filtration membrane (M);

– control system (C);

– power supply (E).

These components are combined using OR and AND logic gates. For instance:

$$\text{Dialysis interruption} = P \cup (C \cap (M \cup L)) \cup E \qquad [4.6]$$

This logical structure can be interpreted using both FTA and BNs, each offering specific advantages depending on the analysis goal (Balakrishnan 2015; Liu et al. 2015).

4.7.2.1. *Fault tree analysis*

The fault tree is a top-down modeling approach. Logical relationships are defined as:

– *OR gate*: the system fails if at least one of the components fails:

$$P_{OR} = 1 - \prod_{i=1}^{n}(1 - F_i(t)) \qquad [4.7]$$

– *AND gate*: the system fails only if all components fail:

$$P_{AND} = 1 - \prod_{i=1}^{n} F_i(t) \qquad [4.8]$$

The overall system failure probability is derived through hierarchical composition of these logic gates.

Advantages are as follows:

– simple and intuitive structure;

– enables qualitative analysis (e.g. identification of critical paths);

– easy to build for linear systems (Balakrishnan 2015).

Limitations are as follows:

– strong independence assumptions between components;

– poor handling of loops and cross-dependencies;

– static model: hard to update when observations are available (Liu et al. 2015).

Applying FTA to Figure 4.2 provides an initial estimation of the dialysis interruption risk. However, it becomes less effective when modeling complex or dynamic relationships – this is where BNs become useful.

4.7.2.2. *BN modeling*

BNs enable causal and probabilistic modeling:

– each node represents a random variable (P, L, M, C, E);

– arcs represent conditional dependencies;

– the graph is directed and acyclic.

The joint probability is factorized as follows:

$$P(P,M,L,C,E,I) =$$

$$P(E) \cdot P(C \mid E) \cdot P(M) \cdot P(L \mid C) \cdot P(P \mid C) \cdot P(I \mid P,M,L,C,E) \qquad [4.9]$$

Each node is associated with a conditional probability table (CPT), which may be defined from a Weibull law:

$$P(\,Xi = 1 \mid \text{Parents}(Xi)\,) = 1 - \exp\left(-\frac{t}{\eta_i}\right)^{\beta_i} \qquad [4.10]$$

Advantages are as follows:

– explicit representation of dependencies;

– capable of integrating empirical or expert knowledge;

– supports inference and real-time updates (Bayesian algorithms) (Hunte et al. 2024; Tonelli et al. 2023).

Limitations are as follows:

– more complex construction process;

– requires richer statistical data;

– computationally intensive (Tonelli et al. 2023).

In the context of Figure 4.3, a BN provides a more realistic modeling of the system behavior, especially for conditional relationships not expressible through FTA.

4.7.2.3. *Comparative analysis and complementarity*

Both methods are complementary (Table 4.3):

– FTA offers a structured overview of the system and quickly highlights critical paths (Balakrishnan 2015).

– BN goes deeper into dynamic probabilistic analysis, accounting for updates and interdependencies (Hunte et al. 2024).

In practice, a fault tree can be transformed into a BN, supporting hybrid analysis. This is particularly relevant in predictive maintenance of biomedical systems, where decisions must be quick, reliable and adaptive (Tonelli et al. 2023).

Criterion	Fault tree analysis (FTA)	Bayesian network (BN)
Structure	Top-down logical	Causal, directed
Dependencies	Assumed independent	Modeled explicitly
Updates	Not well-suited	Enabled through observations
Real-time inference	No	Yes
Complexity	Low	Moderate to high

Table 4.3. *Comparison between fault tree analysis and Bayesian network*

This complementarity justifies the increasing adoption of hybrid diagnostic frameworks combining both models' strengths, especially in critical applications like dialysis systems.

Modeling assumptions are as follows:

– Components are modeled as binary variables (functional/faulty).

– Weibull parameters are calibrated from maintenance or test data.

– Relationships (in BN) are assumed to be time-invariant during analysis.

– Human and environmental factors are excluded for simplification.

4.8. Conclusion

This chapter has established the essential theoretical and methodological foundations for the in-depth analysis of reliability, maintainability and availability (RMA) in dialysis machines. We have systematically presented and explained the analytical tools that will be employed in the subsequent chapter to evaluate and optimize the performance of these critical medical devices. The Weibull distribution, characterized by its flexibility through shape (β) and scale (η) parameters, serves as the cornerstone of this approach. It enables accurate modeling of various failure regimes – including early-life, random and wear-out failures – to which dialysis system components are subjected under real-world operational conditions.

Our integrated methodology combines three complementary approaches, offering a holistic perspective on risk management. Failure mode, effects and criticality analysis (FMEA) provides a systematic bottom-up framework for identifying and prioritizing risks at the component level. FTA contributes a top-down approach for modeling complex logical relationships between critical events and their root causes. Finally, BNs introduce a sophisticated probabilistic framework capable of dynamic analysis that incorporates uncertainty and real-time observational data.

The significance of this methodological framework lies in its ability to transcend traditional approaches often constrained by oversimplifying assumptions. Specifically, the implementation of the Weibull distribution overcomes limitations inherent in exponential models, which presume constant failure rates – an assumption rarely valid for complex systems such as dialysis machines. The complementary nature of these methodologies addresses their individual limitations: while FMEA struggles to capture failure interactions, FTA effectively models these relationships; however, FTA remains static, and BNs introduce essential dynamic capabilities for modern maintenance management.

The theoretical developments presented herein establish the necessary foundation for quantitative modeling of key performance indicators, including MTBF, MTTR and operational availability. The mathematical formalism associated with the Weibull distribution – encompassing reliability R(t), maintainability M(t) and failure rate λ(t) functions – provides the computational framework that will be implemented in the following chapter for applied analysis of dialysis systems.

This preparatory work enables field applications that extend beyond basic diagnostic capabilities toward predictive and optimized equipment management. The integration of these methodological tools will facilitate the development of risk-based maintenance strategies, optimized intervention intervals, streamlined spare parts management and, ultimately, maximized machine availability for patients – all while respecting the budgetary constraints of healthcare institutions.

The subsequent chapter will build upon these theoretical foundations to conduct a comprehensive applied analysis of dialysis machines. This will range from estimating Weibull parameters for critical subsystems to developing optimized, context-specific maintenance plans. This transition from theoretical framework to practical implementation will validate the operational relevance of the deployed methodological tools and demonstrate their concrete value in enhancing care quality and safety for patients with renal failure.

5

Reliability, Maintainability and Availability of the Dialysis Machines based on the Failure History Through Weibull Distribution

5.1. Introduction

In this chapter, we present a numerical study of the dialysis system failure statistics. The working principle and different parts of the dialysis machine are explored and discussed. The statistical count of all failures detected on each dialysis machine is determined and included in the system analysis. The Pareto analysis method will be introduced to determine the priority of the maintenance department interventions by identifying the most serious failures that affect the operational safety of the machines. The Weibull distribution will be used to study the reliability of the hemodialysis system. The Weibull parameters are identified and the efficiency of the chosen distribution validated through the Kolmogorov–Smirnov test. Among the contributions of this work, the maintainability and availability of the dialysis system will be evaluated. They can be effective tools for the assessment of the current situation and for the prediction of the reliability of the upgrade of the operational management of the dialysis system. In addition, the question of the systematic inspection period is treated. At the end of this chapter, a plan of preventive maintenance will be developed with the objective of planning the different types of interventions to guarantee maximum reliability and availability of the dialysis machines.

5.2. Overview of dialysis machine failure history

The considered dialysis system is composed of ten hemodialysis machines (M1, M2, . . ., M10), Fresinus mark and 40088 type. Their commissioning date is August 11, 2012. The dialysis system operates with an average of about 3.5 h per working day, six days a week and usually stops on Sundays. The failure history of the ten devices was documented by the hospital's technical department over the period from 2013 to 2022. The dataset includes the detected failures and their nature, the corrective actions implemented, the associated downtime and the precise time of each breakdown. The time between failures (TBF) for a component is defined as the interval from when the equipment resumes operation following a failure or maintenance event until the next failure occurs. Conversely, the time to repair (TTR) is defined as the duration from the moment a component fails until it is restored to operational status. Both TBF and TTR were calculated in hours and recorded for all ten machines.

The maintenance policy currently applied to the dialysis machines is both preventive and corrective; preventive maintenance is planified and applied periodically, while corrective maintenance is unplanified and performed whenever a fault is detected; the procedure needs emergency intervention by maintenance personnel to return the device to operating condition. Preventive maintenance is performed every 2,500 h of operation. Table 5.1 (from (a) to (j)) illustrates the failure history of all machines, including the date and nature of each failure. The corresponding TBF and TTR are calculated and reported for each intervention. Preventive maintenance includes revisions every 2,500 h as well as technical security checks (TSC).

The tasks of TSC for a dialysis machine involve a series of rigorous technical verifications and safety tests performed before each use. This includes checking pressure systems, verifying circuit integrity, testing safety alarms, calibrating sensors and validating the sterilization of the blood circuit. These operations ensure the safe operation of the device and prevent any risk to the patient during treatment.

Inventory No.: Machine 1			Serial No.: 1V5ATF44		
Failures			Interventions		
Date	Description	TBF (h)	Date	TTR (h)	Description
05/06/2015	Revision 2,500 h	2,450	07/06/2015	7	2,500 h Revision kit change
09/11/2015	Clogged venous pressure sensor	427	11/11/2015	7	Change of hydrophate filter and venous pressure nozzle
20/11/2017	Revision 5,000 h	2,047	29/11/2017	28	5,000 h Revision kit change
27/07/2019	Flow alarm	1,673	31/07/2019	14	Engine check and cleaning Flow calibration
06/03/2020	(TSC)	595	21/04/2020	157.5	Calibration of parameters
16/09/2020	Revision 7,500 h	539	16/09/2020	3.5	7,500 h Revision kit change
05/09/2021	Water leak during disinfection cycle	1,372	07/09/2021	7	Change the bicarbonate pipette seal
01/11/2021	Flow alarm	147	06/11/2021	17.5	Change solenoid valve 34, calibration of parameters

Table 5.1. *(a) Failure history of the hemodialysis machine 1*

Inventory No.: Machine 2			Serial No.: 1V5ATG37		
Failures			Interventions		
Date	Description	TBF (h)	Date	TTR (h)	Description
05/06/2015	Revision 2,500 h	2,450	07/06/2015	7	2,500 h Revision kit change
20/11/2017	Revision 5,000 h	2,481.5	29/11/2017	31.5	5,000 h Revision kit change
23/01/2019	Acid pump problem	1155	24/01/2019	3.5	Acid pump check and cleaning
25/03/2019	Acid pump problem	169	27/03/2019	7	Acid pump trouble shooting: device under test
06/03/2020	TSC and revision 7,500 h	1,039.5	21/04/2020	157.5	Calibration of parameters and change of 7,500 h revision kit

Table 5.1. *(b) Failure history of the hemodialysis machine 2*

Inventory No.: Machine 3			Serial No.: NS1V5ATF77		
Failures			Interventions		
Date	Description	TBF (h)	Date	TTR (h)	Description
29/01/2014	Temperature problem	1,085	06/02/2014	28	Replacing the temperature sensor
04/01/2016	Revision 2,500 h	1,904	13/01/2016	31.5	change of 2,500 h revision kit
19/10/2016	Flow alarm	609	21/10/2016	7	Changing valve 24
24/10/2016	Flow alarm	10.5	25/10/2016	7	Cleaning the hydraulic circuit
29/05/2017	High conductivity	605.5	31/05/2017	7	Conductivity calibration
20/11/2017	Revision 5,000 h	532	20/11/2017	3.5	Change of 5,000 h revision kit
06/03/2020	(TSC)	2,792	22/04/2020	157.5	Calibration of parameters
12/11/2020	Acid pump problem	392	14/11/2020	7	Maintenance and leak check/flow calibration
02/12/2020	Revision 7,500 h	77	02/12/2020	3.5	Change of 7,500 h revision kit

Table 5.1. *(c) Failure history of the hemodialysis machine 3*

Inventory No.: Machine 4			**Serial No.: NS1V5ATG46**		
Failures			**Interventions**		
Date	**Description**	**TBF (h)**	**Date**	**TTR (h)**	**Description**
22/01/2013	Flow alarm	73.5	31/01/2013	35	Flow calibration
15/07/2013	Venous pressure problem	448	16/07/2013	3.5	Unblocking the bicarbonate pump
01/03/2014	High conductivity	717.5	02/03/2014	3.5	Flow calibration
08/10/2014	Acid pump problem	578	09/10/2014	3.5	Problem with the pipes
05/06/2015	Revision 2,500 h	3,650.5	07/06/2018	3.5	2,500 h Revision kit change
29/05/2017	High conductivity Flow alarm	192.5	30/05/2017	3.5	Conductivity and flow calibration
07/08/2017	Flow alarm	553	13/08/2017	24.5	Addition of 2 pipette filters and calibration
12/11/2017	Acid pump problem	168	14/11/2017	7	Maintenance and leak check/flow calibration
28/02/2018	Revision 5,000 h	73.5	10/04/2018	126	5,000 h Revision kit change
07/05/2018	Venous pressure problem	213.5	08/05/2018	3.5	Change of vein nozzle and hydrophobic filter
24/07/2018	Flow alarm	245	25/07/2018	7	Change of vein nozzle and hydrophobic filter
03/08/2018	Acid pump problem	28	04/08/2018	3.5	Change restriction filter and parameter calibration
19/11/2018	Non-absorption of acid pump	224	20/11/2018	3.5	Clamping flange (water leak)
18/09/2019	Venous pressure problem	829.5	19/09/2019	3.5	Flow calibration
06/03/2020	TSC	462	22/04/2020	164.5	Calibration of parameters
02/12/2020	Revision 7,500 h	560	02/12/2020	3.5	7,500 h Revision kit change

Table 5.1. *(d) Failure history of the hemodialysis machine 4*

Inventory No.: Machine 5			Serial No.: 1V5ATG31		
Failures			Interventions		
Date	Description	TBF (h)	Date	TTR (h)	Description
04/01/2016	Revision 2,500 h	3,034.5	13/01/2016	21	2,500 h Revision kit change
29/05/2017	High conductivity	1,389.5	31/05/2017	10.5	Conductivity Calibration
12/06/2017	Flow alarm	42	14/06/2017	10.5	Change of pipette filter and restriction filter
20/11/2017	Revision 5,000 h	441	06/12/2017	42	5,000 h Revision kit change
15/10/2018	Venous pressure problem	871.5	17/10/2018	7	Change of vein nozzle and hydrophobic filter
11/09/2019	Flow alarm	903	11/09/2019	3.5	Flow calibration
06/03/2020	(TSC) and revision 7,500 h	486.5	21/04/2020	157.5	Parameters calibration/7,500 h Revision kit change
29/03/2021	Water leak alarm	896	31/03/2021	7	Checking for leaks during the disinfection cycle
22/07/2021	Low conductivity	304.5	21/08/2021	108.5	Acid pump change

Table 5.1. *(e) Failure history of the hemodialysis machine 5*

Inventory No.: Machine 6			Serial No.: 1V5ATF82		
Failures			Interventions		
Date	Description	TBF (h)	Date	TTR (h)	Description
05/06/2015	Revision 2,500 h	2,450	07/06/2015	7	2,500 h Revision kit change
29/05/2017	High conductivity	2,040.5	31/05/2017	7	Conductivity calibration
07/08/2017	Conductivity problem	192.5	09/08/2017	7	Change of bicarbonate spring kit
28/02/2018	Revision 5,000 h	549.5	10/04/2018	126	5,000 h Revision kit change
03/08/2018	Flow alarm	311.5	07/08/2018	14	Change of restriction filter, flow meter and E24 solenoid valve. Calibration of parameters
06/03/2020	TSC	584.5	22/04/2020	157	Calibration of parameters
03/08/2020	Revision 7,500 h	280	03/08/2020	3.5	7,500 h Revision kit change

Table 5.1. *(f) Failure history of the hemodialysis machine 6*

Inventory No.: Machine 7			Serial No.: 1V5ATG43		
Failures			Interventions		
Date	Description	TBF (h)	Date	TTR (h)	Description
05/06/2015	Revision 2,500 h	2,450	07/06/2015	7	2,500 h Revision kit change
15/05/2017	Pressure sensor problem	1,949.5	17/05/2017	7	Change of hydrophobic filter and vein spinner
20/11/2017	Revision 5,000 h	514.5	06/12/2017	42	5,000 h Revision kit change
06/03/2020	TSC	2,268	28/04/2020	168	Calibration of parameters
18/09/2020	Water leak and Revision 7,500 h	385	21/09/2020	10.5	Elimination of water leaks during pressure testing/7,500 h Revision kit change
16/09/2021	Acid pump alarm	990.5	18/09/2021	7	Elimination of leaks in the hydraulic circuit

Table 5.1. *(g) Failure history of the hemodialysis machine 7*

Inventory No.: Machine 8			Serial No.: 1V5ATG39		
Failures			Interventions		
Date	Description	TBF (h)	Date	TTR (h)	Description
13/03/2015	Acid pump fault alarm	2,226	25/03/2015	45.5	Changing the bicarbonate pump spring kit
04/01/2016	Revision 2,500 h	766.5	13/01/2016	31.5	2,500 h Revision kit change
13/01/2016	Venous pressure problem	3.5	13/01/2016	3.5	Changing the vein nozzle
22/07/2016	Flow alarm	535.5	27/07/2016	14	Cleaning the pump head
19/09/2016	Flow alarm	140	21/09/2016	3.5	Cleaning the pump head
02/08/2017	Flow alarm	878.5	04/08/2017	3.5	2 bar charging valve change
08/01/2018	Conductivity problem	343	09/01/2018	3.5	Conductivity calibration
05/02/2018	Venous pressure problem	66.5	06/02/2018	3.5	Change of hydrophobic filter and vein toupee
28/02/2018	Revision 5,000 h	63	10/04/2018	126	5,000 h Revision kit change
10/08/2018	Flow alarm	336	13/08/2018	10.5	Change of double-way solenoid valve E24 and calibration of parameters
06/03/2020	(TSC)	1,575	28/04/2020	168	Calibration of parameters
08/07/2020	Flow alarm and revision 7,500 h	203	08/07/2020	3.5	Troubleshooting the flow pump/7,500 h Revision kit change
24/02/2021	High flow alarm	633.5	26/02/2021	7	Pump verification and parameter calibration
05/11/2021	Acid pump problem	703.5	16/11/2021	42	Total pump maintenance

Table 5.1. *(h) Failure history of the hemodialysis machine 8*

Inventory No.: Machine 9			Serial No.: 1V5ATG29		
Failures			Interventions		
Date	Description	TBF (h)	Date	TTR (h)	Description
17/04/2015	Pressure sensor problem	2,313.5	22/04/2015	17.5	Change of hydrophobic filter and venous pressure spinner
05/06/2015	Revision 2,500 h	119	07/06/2015	7	2,500 h Revision kit change
29/05/2017	High conductivity	1,988	31/05/2017	7	Conductivity calibration
06/12/2017	Revision 5,000 h	525	06/12/2017	3.5	5,000 h Revision kit change
06/03/2020	(TSC)	3,276	28/04/2020	168	Calibration of parameters
24/06/2020	Revision 7,500 h	154	24/06/2017	3.5	7,500 h Revision kit change

Table 5.1. *(i) Failure history of the hemodialysis machine 9*

Inventory No.: Machine 10			Serial No.: 1V5ATF70		
Failures			Interventions		
Date	Description	TBF (h)	Date	TTR (h)	Description
29/05/2013	Bicarbonate suction problem	413	01/06/2013	7	Changing the bicarbonate suction tube
12/07/2013	Flow alarm	101.5	23/07/2013	42	Flow calibration
01/09/2014	Acid pump problem	1,130.5	03/09/2014	3.5	Changing the hydrophobic filter pressure nozzle
05/06/2015	Revision 2,500 h	763	07/06/2015	7	2,500 h Revision kit change
09/11/2015	Venous pressure problem	427	15/11/2015	17.5	Change hydrophobic filter and venous pressure nozzle
07/08/2017	Water leak	1,750	08/08/2017	3.5	Change 2 concentrate filters, 2 concentrate seals and 2 pipette filters
31/10/2018	Conductivity problem	1,228.5	02/11/2018	7	Calibration of parameters, conductivity and flow rate
09/08/2019	Flow alarm	780.5	16/08/2019	28	Change restriction filter and flow calibration
06/03/2020	TSC	560	28/04/2020	168	Calibration of parameters
29/07/2020	Revision 5,000 h	255.5	29/07/2020	3.5	5,000 h Revision kit change
07/01/2022	Water leak	1,456	07/01/2022	3.5	Changing the acid pipette seal

Table 5.1. *(j) Failure history of the hemodialysis machine 10*

5.3. Pareto analysis method

A Pareto analysis was conducted to identify the hemodialysis machines contributing most significantly to the total number of failures. Rather than focusing on specific failure modes, this approach aimed to reveal machines with disproportionately high failure rates, thus allowing maintenance efforts to be prioritized. Although this method does not directly explain the causes of fleet inhomogeneity, it provides a useful overview of reliability disparities and serves as a basis for further investigations.

The goal is to analyze a phenomenon by representing it by a graph that allows us to determine the existence of a relationship between two groups of data. The Pareto curve, also called the ABC curve, allows us to classify events according to the stake they represent. It also allows us to quickly visualize the priorities of action, make a choice and concentrate on the problems to be treated as a priority. It is a simple way to classify phenomena in order of importance. The interest of the ABC curve is to show that it is more profitable to tackle the resolution of the main causes of defects than to waste precious time treating causes that only rarely appear. It thus allows us to easily distinguish the essential from the secondary.

Machine / Failure	M1		M2		M3		M4		M5	
	TTR	N	TTR	N	TTR	N	TTR	N	TTR	N
Flow parameter calibration problem	**31.5**	**2**	0	0	**14**	**2**	**66.5**	**3**	**3.5**	**1**
Pipe problem	0	0	0	0	0	0	**3.5**	**1**	0	0
Filter problem	0	0	0	0	0	0	**3.5**	**1**	**10.5**	**1**
Acid pump problem	0	0	**10.5**	**2**	**7**	**1**	**10.5**	**2**	0	0
Conductivity problem	0	0	0	0	**7**	**1**	7	**2**	**119**	**2**
Venous pressure nozzle problem	**7**	**1**	0	0	0	0	**10.5**	**3**	**7**	**1**
Water leak problem	**7**	**1**	0	0	0	0	0	**0**	**7**	**1**
Temperature problem	0	0	0	0	**28**	**1**	0	0	0	0
Total TTR and failures	**45.5**	**4**	**10.5**	**2**	**56**	**5**	**101.5**	**12**	**147**	**6**

Table 5.2. *(a) TTR depending on the failure nature of the machines M1 to M5*

Failure \ Machine	M6		M7		M8		M9		M10	
	TTR	N	TTR	N	TTR	N	TTR	N	TTR	N
Flow parameter calibration problem	14	1	0	0	14	3	0	0	70	2
Acid pump problem	0	0	7	1	87.5	2	0	0	3.5	1
Bicarbonate pump problem	0	0	0	0	24.5	3	0	0	7	1
Conductivity problem	14	2	0	0	3.5	1	7	1	3.5	1
Filter problem	0	0	0	0	3.5	1	0	0	0	0
Venous pressure nozzle problem	0	0	7	1	3.5	1	17.5	1	17.5	1
Water leak problem	0	0	10.5	1	0	0	0	0	7	2
Total TTR and failures	28	3	24.5	3	136.5	11	24.5	2	108.5	8

Table 5.2. *(b) TTR depending on the failure nature of the machines M6 to M10*

The Pareto diagram therefore allows us to justify a quality improvement action.

Tables 5.2(a) and 5.2(b) allow us to list the TTR according to the nature of the failures for all dialysis machines. The failure history of the ten machines made it possible to determine several types of failures.

According to the data in Tables 5.2(a) and 5.2(b), it can be seen that machines M4, M5, M8 and M10 are the most affected by failures considering the total number of failures and the total TTR of each machine.

Machine M4							
No.	**Causes**	**TTR (h)**	**Cumul TTR**	**% TTR**	**Nbre cause**	**Cumul cause**	**% Cumul cause**
1	**Flow parameter calibration problem**	66.5	66.5	65.51	3	3	25
2	**Acid pump problem**	10.5	77	75.86	2	5	41.66
3	**Venous pressure nozzle problem**	10.5	87.5	86.20	3	8	66.66
4	**Conductivity problem**	7	94.5	93.10	2	10	83.33
5	**Pipe problem**	3.5	98	96.83	1	11	91.66
6	**Filter problem**	3.5	101.5	100	1	12	100
Machine M5							
No.	**Causes**	**TTR (h)**	**Cumul TTR**	**% TTR**	**Nbre cause**	**Cumul cause**	**% Cumul cause**
1	**Conductivity problem**	119	119	80.95	2	2	33.33
2	**Filter problem**	10.5	129.5	88.09	1	3	50
3	**Venous pressure nozzle problem**	7	136.5	92.85	1	4	66.66
4	**Water leak problem**	7	143.5	97.61	1	5	83.33
5	**Flow parameter calibration problem**	3.5	147	100	1	6	100
Machine M8							
No.	**Causes**	**TTR (h)**	**Cumul TTR**	**% TTR**	**Nbre cause**	**Cumul cause**	**% Cumul cause**
1	**Acid pump problem**	87.5	87.5	64.10	2	2	18.18
2	**Bicarbonate pump problem**	24.5	112	82.05	3	5	45.45
3	**Flow parameter calibration problem**	14	126	92.30	3	8	72.72
4	**Conductivity problem**	3.5	129.5	94.87	1	9	81.81
5	**Filter problem**	3.5	133	97.43	1	10	90.90
6	**Venous pressure nozzle problem**	3.5	136.5	100	1	11	100

Machine M10							
No.	**Causes**	**TTR (h)**	**Cumul TTR**	**% TTR**	**Nbre cause**	**Cumul cause**	**% Cumul cause**
1	**Flow parameter calibration problem**	70	70	64.51	2	2	25
2	**Venous pressure nozzle problem**	17.5	87.5	80.64	1	3	37.5
3	**Bicarbonate pump problem**	7	94.5	87.09	1	4	50
6	**Water leak problem**	7	101.5	93.54	2	6	75
4	**Acid pump problem**	3.5	105	96.77	1	7	87.5
5	**Conductivity problem**	3.5	108.5	100	1	8	100

Table 5.3. *Cumulative causes of failures for hemodialysis machines*

In the following paragraph, a Pareto analysis of the machines M4, M5, M8, and M10 will be performed to determine the most dangerous faults and their influences, which allows developing an effective maintenance plan. To apply the Pareto analysis method, we must, first, return to Tables 5.2(a) and 5.2(b) to classify the causes in descending order according to the hours of breakdowns (TTR), which allows us to build Table 5.3, which summarizes the causes of failures in various groups of causes. Then, we will proceed to construct the Pareto curve. According to the Pareto analysis presented in Figure 5.1, it can be seen that for machines M4, M5, M8 and M10, 70% of the failures are caused by 30% of the causes (zone A). Therefore, the maintenance department must focus its interventions on eliminating these 30% of the causes as a priority to minimize downtime. The second zone (zone B) is delimited between 30% and 70% of the cumulative causes, which gives a cumulative of 90% of the breakdowns. This zone is less critical than the first zone A. The third zone (zone C) is defined by the 30% of the remaining causes delimited between 70% and 100% and which caused only 10% of the failures. The causes detected in this zone are negligible and do not present any risk to the machine's operating safety. It should be noted that this Pareto analysis makes it possible to establish a reliable preventive maintenance plan by taking into account the failure history of machine and the severity of the breakdowns encountered.

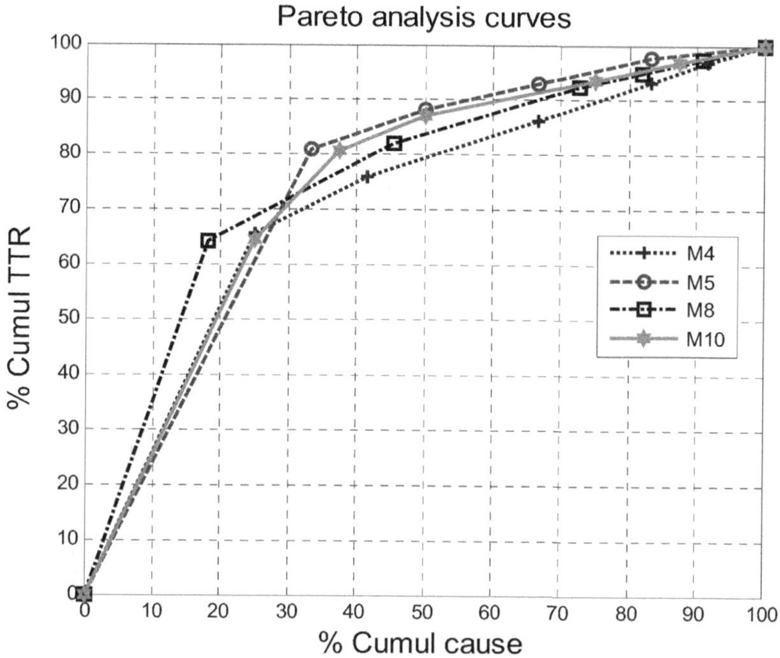

Figure 5.1. *The Pareto analysis curve of hemodialysis machines. For a color version of this figure, see www.iste.co.uk/bouchoucha/optimizing.zip*

5.4. Estimation of the Weibull parameters

Although failure times in repairable systems are generally neither independent nor identically distributed due to maintenance effects and aging, these dependencies were minimized in this study. Only failure-induced corrective actions were considered, excluding maintenance interventions and their durations, such as preventive tasks and associated downtime. This allowed the analysis to focus solely on time intervals between actual failures. Assuming that each repair restores the system to an "as good as new" condition, the independence and identical distribution of failure times becomes a reasonable approximation. Consequently, the Weibull distribution was selected and validated using goodness-of-fit criteria, ensuring its suitability for the observed failure patterns.

The Weibull model is widely used in reliability analysis for its flexibility and interpretability. In this work, the two-parameter Weibull distribution was adopted to describe the time-to-failure behavior of hemodialysis machines. Despite some machines (e.g. M1, M2, M5 and M9) showing high initial reliability with first failures after 2,000 h, the two-parameter model remains appropriate. Its simplicity and robustness are particularly useful when individual machine datasets are limited, allowing consistent modeling across the fleet. The shape and scale parameters effectively describe the failure trend and characteristic life. Although a three-parameter version could theoretically capture an initial failure-free period, it introduces additional estimation complexity and may lead to overfitting, especially with small sample sizes. For these reasons, the two-parameter Weibull model was consistently used throughout the analysis.

Among the principal estimation techniques are the maximum likelihood estimation (MLE), the least squares method (LSM) and the method of moments (MOM). MLE is generally regarded as the most statistically efficient method, particularly for large samples or censored data, although it requires iterative computations and is sensitive to initial parameter values (Ferreira and Silva 2017). LSM, based on linear regression of transformed data, is computationally simpler but may produce biased results in the presence of data variability or censoring (David and Jahnke 2004). MOM, which uses the sample's mean and variance to estimate parameters, is easy to implement and useful for quick approximations, although it is generally less accurate than MLE (Ianculescu and Anghel 2025). Graphical methods are also employed, particularly for exploratory analyses, but they lack the precision and reproducibility of analytical approaches (Shahriari et al. 2024). Comparative studies have shown that MLE offers superior performance in most reliability contexts, while LSM and MOM remain acceptable alternatives when computational resources or data quality are limited (Frolov et al. 2024).

The Weibull probability plot method was chosen for this study due to its simplicity, visual clarity and effectiveness in identifying the underlying distribution of failure data. This graphical method allows an intuitive assessment of the fit by plotting ordered failure times against cumulative probability on a logarithmic scale. If the data follows a Weibull distribution, the points should align along a straight line, making it easier to spot outliers or deviations. Additionally, the Weibull probability plot enables a straightforward visual evaluation of the shape and scale parameters, which is especially useful for small sample sizes. While methods like maximum likelihood or method of moments are more precise and suited for larger datasets, the Weibull probability plot offers a balance of accuracy and simplicity, making it a practical choice in this study's context.

The cumulative density function F(t) is given through Weibull distribution as:

$$F(t) = 1 - \exp\left(-\frac{t}{\eta}\right)^{\beta} \quad [5.1]$$

In the goal to determine the Weibull parameters (β and η), the linearization of the cumulative density function is used and determined using Weibull distribution as follows:

$$F(t) - 1 = -\exp\left(-\frac{t}{\eta}\right)^{\beta} \quad [5.2]$$

$$\ln\big(1 - F(t)\big) = -\left(\frac{t}{\eta}\right)^{\beta} \quad [5.3]$$

$$\ln\big(-\ln\big(1 - F(t)\big)\big) = \ln\left(\frac{t}{\eta}\right)^{\beta} \quad [5.4]$$

$$\ln(\ln\left(\frac{1}{1-F(t)}\right) = \beta\ln(t) - \beta\ln(\eta) \quad [5.5]$$

The linearization of the cumulative density function equation leads to establish the behind linear equation with a slop of β and an intercept of $\beta\ln(\eta)$:

$$y = \beta x - \beta\ln(\eta)\text{ , where } y = \ln(\ln\left(\frac{1}{1-F(t)}\right) \text{ and } x = \ln(t) \quad [5.6]$$

Failure times (TBF_i) were collected and arranged in ascending order to estimate the cumulative density function F(ti) using the median rank method (Pasha et al. 2006), as recommended for small sample sizes ($n < 20$). The estimation was performed using Benard's approximation, expressed as:

$$F(t_i) = \frac{i-0.3}{n+0.4} \quad [5.7]$$

where i denotes the rank of the failure and n is the total number of observed failures.

Additional formulas, such as the mean ranking and systematic methods, are also widely used in other domains, depending on the sample size and the specific objectives of the analysis (Ahsan et al. 2019).

Benard's formula is an efficient approximation to the median rank estimator. The Benard's median rank gives a very satisfactory performance and it is the most extensively employed to estimate $F(t_i)$ (Kim and Yum 2008). The complete data ranking method involves listing the time-to-failure data from smallest to largest, then using Benard's formula to affect median ranks to each defect.

To prepare Tables 5.4 ((a)–(d)), the observed times between failures ($TBFi$) for each machine were first extracted from the maintenance records. These values were then arranged in ascending order to facilitate the estimation of the cumulative distribution function $F(t_i)$ using the median rank method. The natural logarithms of both $TBFi$ and the transformed probability $Ln(-Ln(1-F(t_i))$ were then computed. This linearized form allows the use of a Weibull probability plot to estimate the shape and scale parameters of the two-parameter Weibull distribution through linear regression.

The necessary computational steps performed to estimate the Weibull parameters (β and η) of the machines M4, M5, M8 and M10 are given in Tables 5.4((a)–(d)). According to the linear equation of the cumulative density function (Fenina and Bouchoucha 2023), the Weibull probability plot of machines M4, M5, M8 and M10 is illustrated in Figure 5.2.

The Weibull probability plot is employed to examine the precision of the selected distribution as a model for the recorded data, and also to calculate the shape and scale parameters. According to the statistical treatments, it can be confirmed that the data respects a Weibull distribution, which offers the best fit of life data. The choice of distribution depends on the best fit, which allows for assessing the precision of the results (Fenina and Bouchoucha 2023).

The linear equation of the cumulative density function is drawn in Figures 5.2((a)–(d)). The straight line is described by the equations according to Table 5.5. From the study of the equation of the straight line, the Weibull parameters (β and η) can be extracted and given in Table 5.5.

M4				
i	**TBF (h)**	$F(t_i) = \frac{i-0.3}{n+0.4}$	***ln*** **(TBF)**	$ln(-ln(1-F(t_i))$
1	28	0.04191616	3.33220451	–3.150750248
2	73.5	0.10365853	4.29728541	–2.212435171
3	73.5	0.16167664	4.29728541	–1.735276787
4	168	0.22560975	5.12396397	–1.363831223
5	192.5	0.28658536	5.26009615	–1.085619606
6	213.5	0.34756097	5.36363683	–0.850883249
7	224	0.40119760	5.41164605	–0.667823317
8	245	0.46951219	5.68357977	–0.4557721
9	448	0.5304878	6.10479323	–0.279633224
10	462	0.59146341	6.13556489	–0.110737393
11	553	0.65243902	6.315358	0.055259807
12	560	0.71341463	6.32793678	0.222918575
13	578	0.77439024	6.35957387	0.398070165
14	717.5	0.83536585	6.57577295	0.590022796
15	829.5	0.89634146	6.72082311	0.818304317
16	3,650.5	0.95731707	8.20261942	1.148657603

Table 5.4. *(a) The necessary calculation steps of hemodialysis machine M4*

M5				
i	**TBF (h)**	$F(t_i) = \frac{i-0.3}{n+0.4}$	***ln*** **(TBF)**	$ln(-ln(1-F(t_i))$
1	42	0.07446809	3.73766962	–2.558940818
2	304.5	0.18085106	5.71867109	–1.611994375
3	441	0.28723404	6.08904488	–1.082929422
4	486.5	0.39361702	6.1872369	–0.69266027
5	871.5	0.5	6.77021586	–0.366512921
6	896	0.60638298	6.79794041	–0.070018179
7	903	0.71276596	6.80572255	0.221107814
8	1389.5	0.81914894	7.23669925	0.536540994
9	3034.5	0.92553191	8.01780195	0.954505028

Table 5.4. *(b) The necessary calculation steps of hemodialysis machine M5*

M8				
i	TBF (h)	$F(t_i) = \frac{i-0.3}{n+0.4}$	*ln* (TBF)	$ln(-ln(1-F(t_i))$
1	3.5	0.04861111	1.25276297	–2.99909043
2	63	0.11805556	4.14313473	–2.07444434
3	66.5	0.1875	4.19720195	–1.57195252
4	140	0.25694444	4.94164242	–1.21407544
5	203	0.32638889	5.31320598	–0.92861050
6	336	0.39583333	5.81711116	–0.68536716
7	343	0.46527778	5.83773045	–0.46839232
8	535.5	0.53472222	6.28320089	–0.26772170
9	633.5	0.60416667	6.45126	–0.07605845
10	703.5	0.67361111	6.55606788	0.113030157
11	766.5	0.74305556	6.6418347	0.306672154
12	878.8	0.8125	6.77855734	0.515201894
13	1,575	0.88194444	7.36201055	0.75921576
14	2,226	0.95138889	7.70796153	1.106548431

Table 5.4. *(c) The necessary calculation steps of hemodialysis machine M8*

M10				
i	TBF (h)	$F(t_i) = \frac{i-0.3}{n+0.4}$	*ln* (TBF)	$ln(-ln(1-F(t_i))$
1	101.5	0.0614035	4.620058	–2.758770808
2	255.5	0.149122	5.543222	–1.823327725
3	413	0.236842	6.023447	–1.308258602
4	427	0.324561	6.056784	–0.935491323
5	560	0.412280	6.327936	–0.632041114
6	763	0.5	6.637258	–0.366512921
7	780.5	0.587719	6.659934	–0.120980941
8	1,130.5	0.675438	7.030415	0.11803237
9	1,228.5	0.763157	7.113549	0.364894181
10	1,456	0.850877	7.283448	0.643423761
11	1,750	0.938596	7.467371	1.026144924

Table 5.4. *(d) The necessary calculation steps of hemodialysis machine M10*

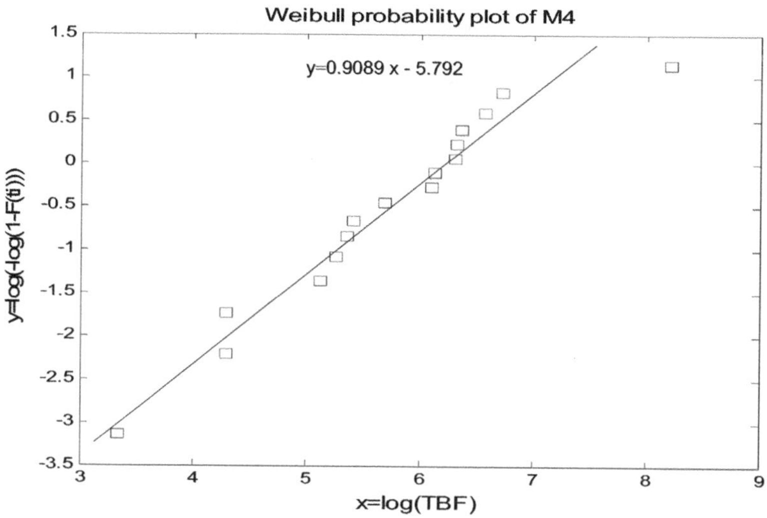

Figure 5.2. *(a) Weibull probability plot of hemodialysis machine M4*

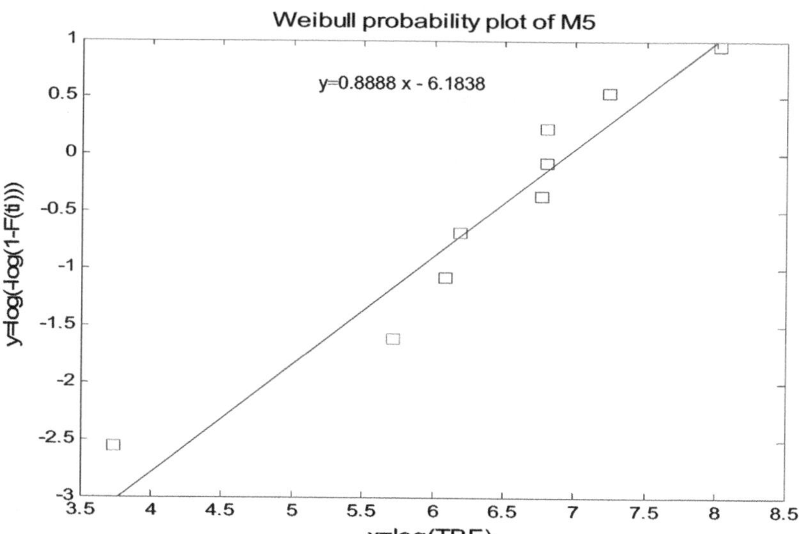

Figure 5.2. *(b) Weibull probability plot of hemodialysis machine M5*

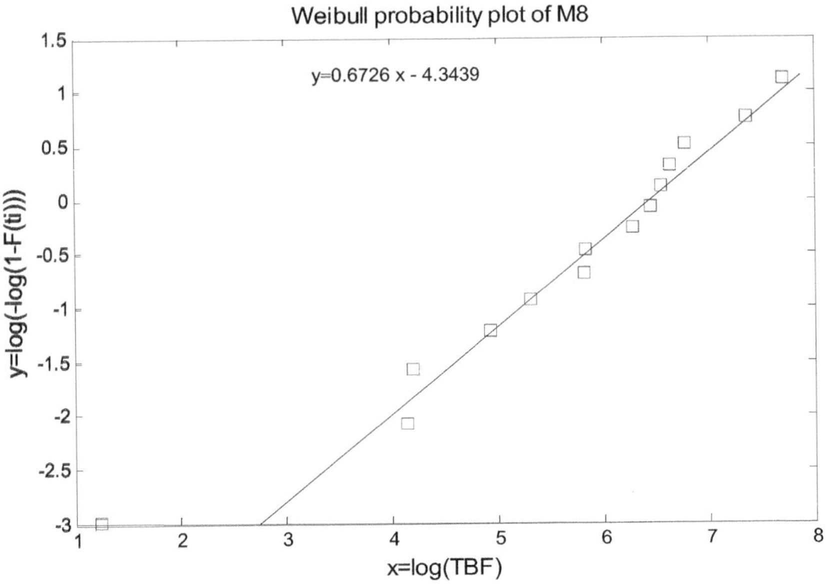

Figure 5.2. *(c) Weibull probability plot of hemodialysis machine M8*

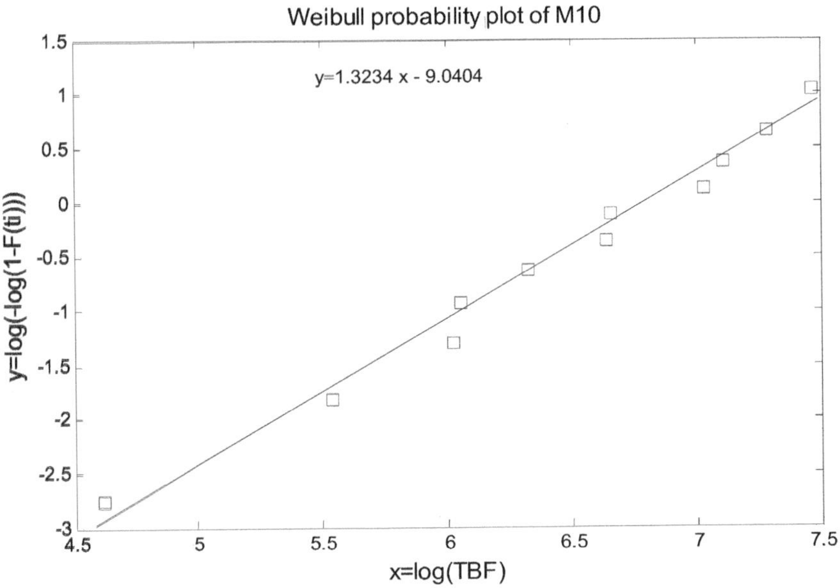

Figure 5.2. *(d) Weibull probability plot of hemodialysis machine M10*

	Linear equation	β	η
M4	$y = 0.9089\,x - 5.792$	0.908	589.25
M5	$y = 0.888\,x - 6.1838$	0.8888	1,056.3
M8	$y = 0.6726\,x - 4.3439$	0.672	640.85
M10	$y = 1.3234\,x - 9.0404$	1.323	927.3

Table 5.5. *Estimation of the Weibull parameters for hemodialysis machines*

5.4.1. *Goodness-of-fit*

In statistics, various techniques are available to assess the goodness-of-fit, such as the Kolmogorov–Smirnov and Anderson–Darling tests. However, in studies of this type, the simple correlation coefficient is often preferred to evaluate how well data fit a straight line (Luko 1999). The correlation coefficient is widely used to measure the strength of a linear relationship between two variables (Khalaf and Hamam 2013). In the analysis of hemodialysis device failure data using the Weibull distribution, the median rank method allows for the determination of the correlation coefficient R^2, providing an indicator of the model's fit (Zou et al. 2003).

The correlation coefficient R^2 is calculated and given in Table 5.6. According to this study, the fit is good for the dialysis machines M4, M5, M8 and M10.

In the context of fitting failure data from hemodialysis devices to a Weibull distribution using the median rank method, the coefficient of determination R^2 proves to be more effective than the simple correlation coefficient R. While both measures assess the strength of a linear relationship, they serve distinct purposes. The coefficient R quantifies the degree of linear correlation between two variables, but it does not directly indicate how well the regression model explains the variability in the data. Conversely, R^2 provides a more meaningful interpretation in the context of goodness-of-fit. Specifically, R^2 represents the proportion of variance in the observed data that is explained by the fitted model. This makes it particularly suitable for evaluating the adequacy of a Weibull probability plot, which is linearized using the median rank method (Krit et al. 2021). In such a linearized Weibull analysis, a high R^2 value directly indicates that the linear fit, and by extension, the Weibull model, closely matches the empirical failure data. Moreover, R^2 is always non-negative and lies between 0 and 1, facilitating straightforward interpretation: values closer to 1 signify a better fit. In contrast, the sign of R may reflect the direction of the relationship but does not affect the quality of the model fit, and $|R|$ alone may be misleading when used to assess fit quality (Schober et al. 2018).

Therefore, in this application, where the goal is to evaluate how well a Weibull distribution fits observed failure data of dialysis machines, R^2 is not only more informative but also more appropriate than R (Fenina et al. 2024a). This is consistent with the reported results, where high R^2 values for machines M4, M5, M8 and M10 confirm the adequacy of the Weibull model in those cases.

	M4	M5	M8	M10
Correlation coefficient R^2	$R^2 = 0.9396$	$R^2 = 0.9278$	$R^2 = 0.9472$	$R^2 = 0.985$

Table 5.6. *The Correlation coefficient of hemodialysis machines*

5.4.2. *Kolmogorov–Smirnov (K–S) test*

Before validating all of the reliability distribution, it is indispensable to check the hypothesis to know if we should admit or reject the model offered by the K–S test with a confidence threshold of $\alpha = 5\%$. This test involves computing the gap between the real function F(t) and the theoretical function F(i), where F(t) is calculated through the Weibull distribution as $F(t) = 1 - \exp\left(-\frac{t}{\eta}\right)^{\beta}$ and F(i) is given by the median rank formula as $F(i) = \frac{i-0.3}{n+0.4}$. Then, take the maximum in absolute value $DNmax = |F(i) - F(t)|$. The value of DNmax is compared with DNα, given in the table of the law of K–S (Appendix 2).

The calculation steps of K–S test for hemodialysis machines are detailed in Tables 5.7((a)–(d)).

According to the table of the law of K–S, $D_{Nmax} < D_{N,\alpha}$ for M4, M5, M8 and M10 (Table 5.8), which means that the Weibull model is validated.

M4				
i	**TBF (h)**	**F(i)**	**F(t)**	**D_{Nmax}**
1	28	0.04191	0.06014	0.0182
2	73.5	0.10365	0.13842	0.0347
3	73.5	0.16167	0.13842	0.0232
4	168	0.2256	0.27066	0.0450
5	192.5	0.28658	0.30033	0.0137
6	213.5	0.34756	0.32453	0.0230
7	224	0.40119	0.33624	0.0649
8	245	0.46951	0.3589	0.1106
9	448	0.53048	0.53654	0.006
10	462	0.591463	0.54653	0.0449
11	553	0.65243	0.60587	0.0465
12	560	0.71341	0.61006	0.1033
13	578	0.77439	0.62061	0.1537
14	717.5	0.83536	0.69254	0.1428
15	829.5	0.89634	0.73958	**0.1567**
16	3,650.5	0.95731	0.99429	0.0369

Table 5.7. *(a) Calculation steps of Kolmogorov–Smirnov (K–S) test for hemodialysis machine M4*

M 5				
i	**TBF (h)**	**F(i)**	**F(t)**	**D_{Nmax}**
1	42	0.07446	0.05224	0.0222
2	304.5	0.18085	0.27201	0.0912
3	441	0.28723	0.35901	0.0718
4	486.5	0.39361	0.38525	0.0084
5	871.5	0.5	0.56576	0.0685
6	896	0.60638	0,57518	0.0312
7	903	0.71276	0.57783	**0.1349**
8	1,389.5	0.81914	0.72676	0.0924
9	3,034.5	0.92553	0.94183	0.0163

Table 5.7. *(b) Calculation steps of Kolmogorov–Smirnov (K–S) test for hemodialysis machine M5*

M8				
i	**TBF (h)**	**F(i)**	**F(t)**	D_{Nmax}
1	3.5	0.0486	0.0157	0.0329
2	63	0.11805	0.1171	0.0009
3	66.5	0.1875	0.1218	0.0657
4	140	0.25694	0.2082	0.0487
5	203	0.32638	0.2725	0.0539
6	336	0.3958	0.3913	0.0045
7	343	0.46527	0.3970	0.0682
8	535.5	0.5347	0.5406	0.0059
9	633.5	0.6041	0.6043	0.00017
10	703.5	0.6736	0.6465	0.0271
11	766.5	0.743	0.6822	0.0608
12	878.8	0.8125	0.7413	0.0712
13	1,575	0.8819	0.9994	**0.1175**
14	2,226	0.95138	1.1213	0.01699

Table 5.7. *(c) Calculation steps of Kolmogorov–Smirnov (K–S) test for hemodialysis machine M8*

M10				
i	**TBF (h)**	**F(i)**	**F(t)**	D_{Nmax}
1	101.5	0.0614	0.0289	0.0325
2	255.5	0.149	0.1022	0.0468
3	413	0.2368	0.2008	0.036
4	427	0.32456	0.2104	0.1141
5	560	0.41228	0.3086	0.1037
6	763	0.5	0.4730	0.027
7	780.5	0.58772	0.4876	0.1001
8	1,130.5	0.6754	0.7733	0.0979
9	1,228.5	0.76315	0.8456	0.0824
10	1,456	0.85	0.9893	0.1393
11	1,750	0.9385	1.1138	**0.1753**

Table 5.7. *(d) Calculation steps of Kolmogorov–Smirnov (K–S) test for hemodialysis machine M10*

	M4	M5	M8	M10
D_{Nmax}	0.1567	0.1349	0.1175	0.1753
$D_{N,\alpha}$	0.349	0.432	0.349	0.391

Table 5.8. *Kolmogorov–Smirnov test results of hemodialysis machines*

5.5. Reliability analysis through Weibull distribution

The reliability of hemodialysis devices is crucial for nephrologists to ensure both patient safety and the effectiveness and consistency of treatment. Inadequate maintenance can lead to suboptimal dialysis sessions and increased operational difficulties. Therefore, implementing an effective maintenance policy is essential to enhance the reliability and performance of hemodialysis machines.

The failure rate is given using the two parameters Weibull distribution β and η (the location parameter γ = 0) by the following equation:

$$\lambda(t) = \frac{\beta}{\eta}\left(\frac{t}{\eta}\right)^{\beta-1} \qquad [5.8]$$

The reliability function is formulated as follows:

$$R(t) = \exp\left(-\left(\frac{t}{\eta}\right)^{\beta}\right) \qquad [5.9]$$

The cumulative density function is known in the following form:

$$F(t) = 1 - \exp\left(-\left(\frac{t}{\eta}\right)^{\beta}\right) = 1 - R(t) \qquad [5.10]$$

The probability density function is expressed as:

$$f(t) = \frac{\beta}{\eta}\left(\frac{t}{\eta}\right)^{\beta-1} \exp\left(-\left(\frac{t}{\eta}\right)^{\beta}\right) = \lambda(t) \times R(t) \qquad [5.11]$$

The mean time between failure (MTBF) is formulated in the following equation:

$$\text{MTBF} = \int_0^{+\infty} R(t) = \eta.\Gamma\left(1 + \frac{1}{\beta}\right) \qquad [5.12]$$

where Γ is the gamma function given in Appendix 1.

In Table 5.9, the MTBF and the reliability parameters (λ(t), R(t), F(t), f(t)) for the hemodialysis machines were presented. The reliability parameters were calculated at t = MTBF. These indicators proved to be useful tools in evaluating the actual situation and in estimating the reliability, particularly, in the short term, for improving the operation strategy of the system. The reliability parameters are calculated, presented and discussed in the following paragraph.

Device	M4	M5	M8	M10
Failure number	14	9	14	11
β	0.908	0.888	0.672	1.323
η	589	1,057	941	928
Life phase	Useful	Useful	Early	Aging
MTBF	620	1,112	876	857
R(t)	0.3508	0.3513142	0.2912589	0.4065484
F(t)	0.6492	0.648685	0.708741	0.593451
λ(t)	0.0015488	0.0008353	0.0009462	0.0013894
f(t)	0.000543	0.000293	0.000275	0.000564

Table 5.9. *Estimation of the reliability indicators of the hemodialysis machines*

The reliability parameters are presented as histograms in Figures 5.3((a)–(d)), providing a clearer interpretation of the data summarized in Table 5.9. Additionally, the functions λ(t), R(t), F(t) and f(t) for the hemodialysis devices are illustrated in Figures 5.4((a)–(d)) as a function of the time between failures (TBFs).

The analysis indicates that the reliability R(t) of machine M10 requires immediate attention (Fenina et al. 2024b), as its failure rate λ(t) increases significantly over time. Machine M10 is in the aging phase ($\beta > 1$) and must be closely monitored. In contrast, the failure rates of machines M4 and M5 remain constant, indicating operation within their useful life period ($\beta \approx 1$). Machine M8, in the early-life stage ($\beta < 1$), exhibits a decreasing failure rate. To mitigate the impact of failures on dialysis treatments, it is recommended to enhance operational policies, with maintenance efforts prioritized for M10 machines currently in the aging phase.

Among the machines analyzed, M4 exhibits a lower MTBFs of 620 h compared to M5 (1,112 h), M8 (876 h) and M10 (857 h). This difference suggests a relatively higher failure frequency for M4. However, this value remains within an acceptable operational range for hemodialysis machines and does not, on its own, justify the implementation of a distinct maintenance strategy. Several contextual factors, such as usage intensity, patient load or environmental conditions, may contribute to this difference. Therefore, while the lower MTBF of M4 is noted, it is not considered sufficient to warrant an immediate adjustment in maintenance planning. Instead, it underscores the need for further investigation into the specific operating context of M4 before proposing targeted actions. This approach ensures that maintenance recommendations remain data-driven and consistent across all machines.

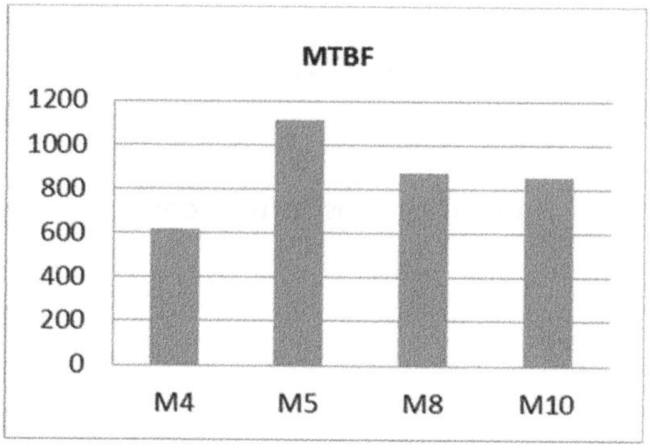

Figure 5.3. *(a) Histogram of the MTBF of hemodialysis machines*

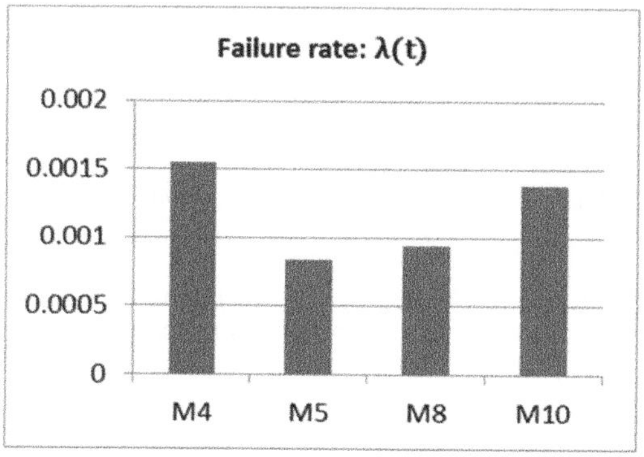

Figure 5.3. *(b) Histogram of the failure rate of hemodialysis machines*

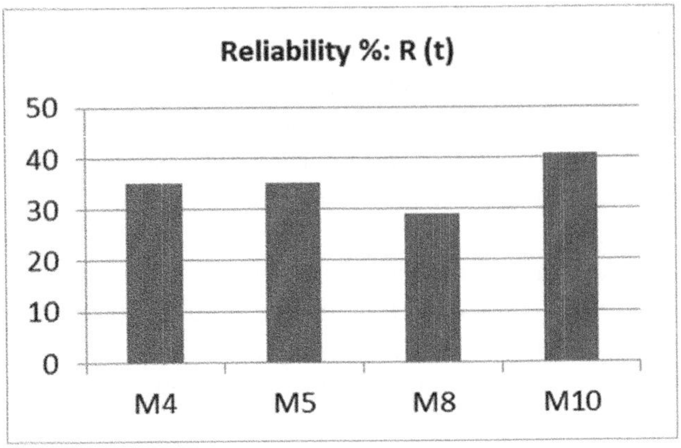

Figure 5.3. *(c) Histogram of the reliability of hemodialysis machines*

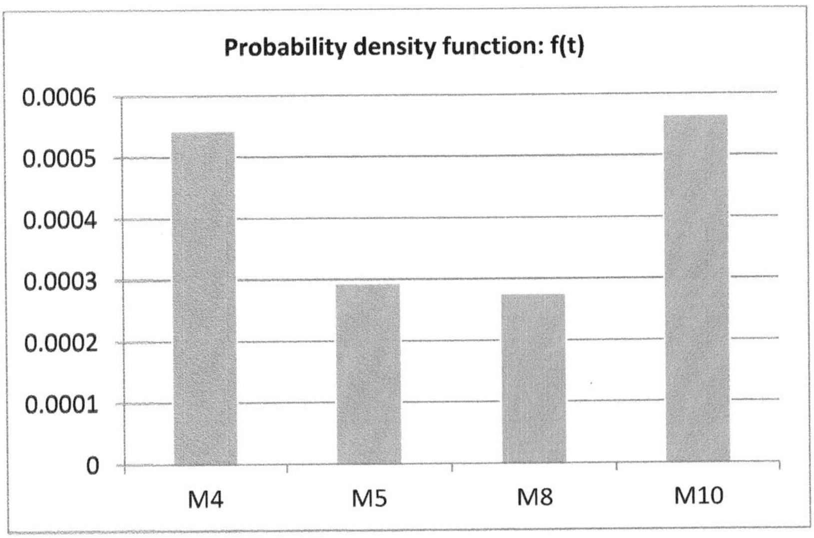

Figure 5.3. *(d) Histogram of the probability density function of hemodialysis machines*

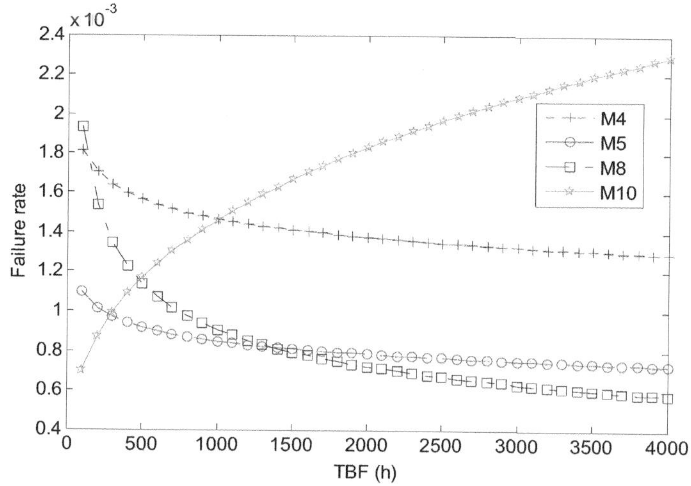

Figure 5.4. *(a) Failure rate of hemodialysis machines. For a color version of this figure, see www.iste.co.uk/bouchoucha/optimizing.zip*

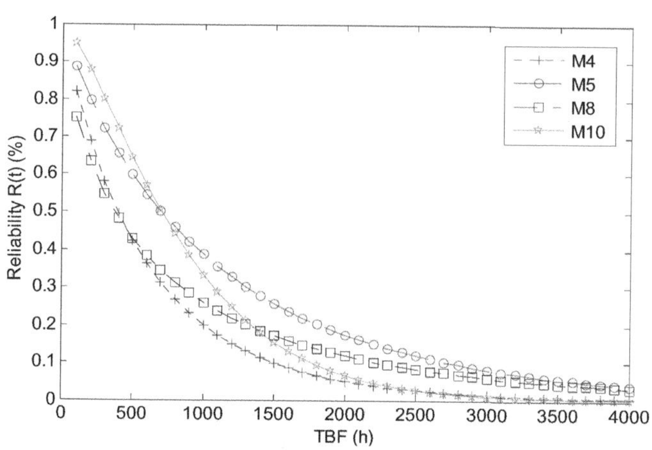

Figure 5.4. *(b) Reliability of hemodialysis machines. For a color version of this figure, see www.iste.co.uk/bouchoucha/optimizing.zip*

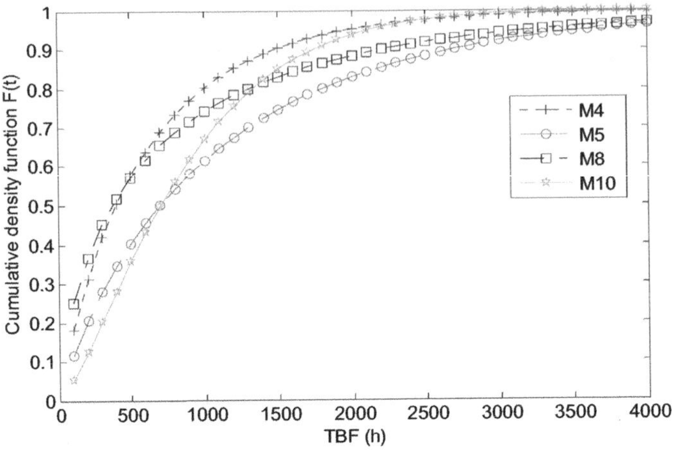

Figure 5.4. *(c) Cumulative density function of hemodialysis machines. For a color version of this figure, see www.iste.co.uk/bouchoucha/optimizing.zip*

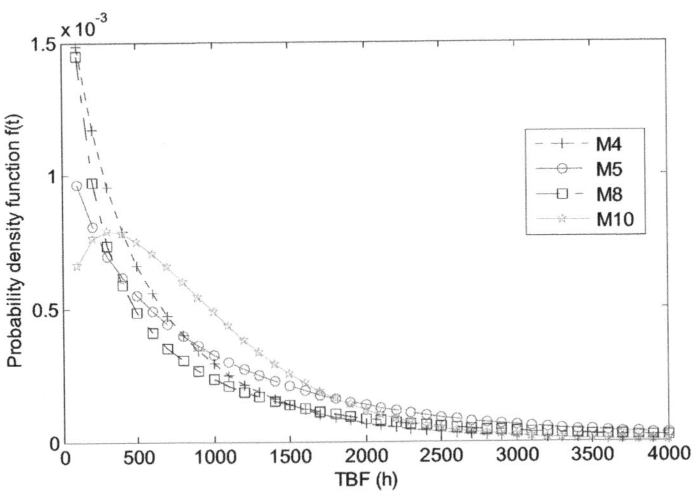

Figure 5.4. *(d) Probability density function of hemodialysis machines. For a color version of this figure, see www.iste.co.uk/bouchoucha/optimizing.zip*

5.6. Maintainability study

The concept of maintainability must consider the characteristics of manufacturing designs and implementations that affect a device's ability to meet specific usage requirements. It is usually measured by the duration of maintenance operations and the requirements for preventive maintenance. In given conditions of use, maintainability is the efficiency of a machine to be maintained or renovated to a state in which it can execute its requested function when maintenance is carried out under given conditions with the prescribed procedures and materials; there is also another probabilistic definition of maintainability, and it is the probability of putting a system into operating condition in a given time with given means and in the conditions of initial reliability. The maintainability of equipment depends essentially on its accessibility and the ease of disassembly and its consecutive elements and their interchangeability.

The maintainability function is written in the following relation:

$$M(t) = 1 - \exp(-\mu t) \quad [5.13]$$

where μ is the repair rate, expressed according the following relationship:

$$\mu = \frac{1}{MTTR} \quad [5.14]$$

where

$$MTTR = \frac{\sum_{i=1}^{N} TTR}{N} \quad [5.15]$$

where TTR is the time to repair and N is the failure number.

Table 5.10 provides the repair rate μ of M4, M5, M8 and M10, extracted from the data presented in Tables 5.2(a) and 5.2(b).

	M4	M5	M8	M10
μ	0.118	0.04	0.08	0.073

Table 5.10. *Repair rate of hemodialysis machines*

Table 5.11 summarizes the calculation of maintainability according to the failure history already mentioned in Table 5.3 and using equation [5.13]. The data shown in Table 5.11 are plotted to obtain the graph in Figure 5.5.

The maintainability functions of machines M4, M5, M8 and M10, plotted as a function of TTR, provide valuable insights into their repair performance (Figure 5.5). The curve of M4 rises sharply and reaches a maintainability close to 1 in less than 50 h, indicating that repairs on this machine are consistently quick and efficient. M8 and M10 also exhibit good maintainability, with similar behavior and fast convergence to full repair probability. In contrast, M5 shows a slower increase in maintainability, requiring over 150 h to approach 1, suggesting that repairs are more variable or time consuming for this unit. These findings reveal that although M4 has a lower MTBFs compared to the others, its excellent maintainability mitigates its higher failure frequency. Conversely, M5, despite its higher MTBF, could suffer from reduced operational availability due to longer or more unpredictable repairs. Therefore, the maintenance strategy should not rely solely on MTBF values but should also account for maintainability performance, which supports the decision to maintain a unified maintenance plan for M4 at this stage.

TTR (h)	M(t)			
	M4	M5	M8	M10
10	0.6927	0.3297	0.5507	0.5181
40	0.9911	0.7981	0.592	0.9461
70	0.9997	0.9392	0.9963	0.994
100	1	0.9817	0.9997	0.9993
150	1	0.9975	1	1
180	1	0.9993	1	1
200	1	0.9997	1	1
230	1	0.9999	1	1
270	1	1	1	1
300	1	1	1	1

Table 5.11. *Maintenability calculation for hemodialysis machines*

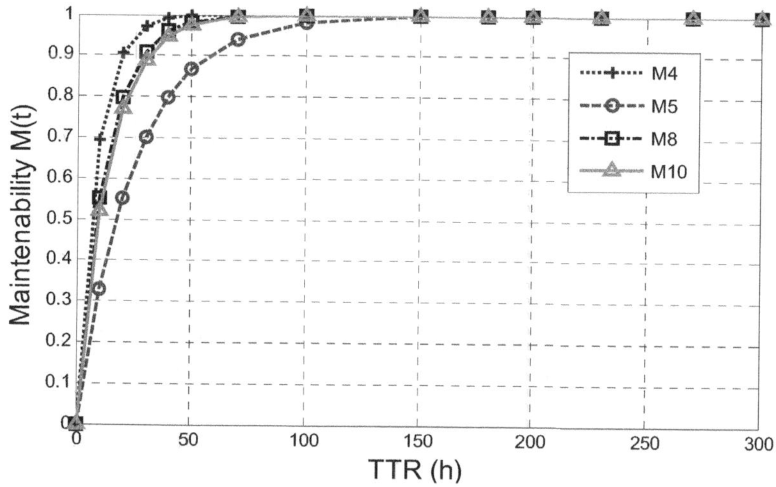

Figure 5.5. *Maintenability of hemodialysis machines. For a color version of this figure, see www.iste.co.uk/bouchoucha/optimizing.zip*

5.7. Availability study

Under the combined aspects of reliability, availability is the probability of a device to be in working order according to prescribed maintenances and for a well-defined period of time. For availability, we must determine both sides of this theoretical and instantaneous parameter, which is directly linked to repair time.

5.7.1. *Theoretical intrinsic availability*

The theoretical intrinsic availability is determined based on the average of good operation and the average of repair. This availability is evaluated by taking into account the average uptime and the average repair time.

It is given by the following relationship:

$$D_i = \frac{MTBF}{MTBF+MTTR} \qquad [5.16]$$

The MTTR, MTBF and D_i of the dialysis machines M4, M5, M8 and M10 are calculated and given in Table 5.12.

	M4	M5	M8	M10
MTTR (h)	8.45	24.5	12.4	13.56
MTBF (h)	620	1,112	876	857
D_i	0.9865	0.978	0.9859	0.9843

Table 5.12. *MTTR, MTBF and D_i of hemodialysis machines*

The value of theoretical availability is satisfactory since it approaches 1. This means that full-time availability requires zero repair time; therefore, there is no breakdown. The value of theoretical intrinsic availability D_i is also linked to the uptime. To improve availability, we must improve the uptime.

5.7.2. *Instantaneous availability*

Instantaneous availability is a function of the failure rate and the repair rate, and is given as follows:

$$D(t) = \frac{\mu}{\mu+\lambda} + \frac{\lambda}{\mu+\lambda} exp(-(\mu+\lambda)t) \qquad [5.17]$$

where λ is the failure rate given as:

$$\lambda = \frac{1}{MTBF} \qquad [5.18]$$

Table 5.13 summarizes the calculation of instantaneous availability through the equation [5.17]. The data given in Table 5.13 are plotted in Figure 5.6.

The availability curves of machines M4, M5, M8 and M10, plotted as a function of the repair time (TTR), reveal important insights into their operational performance (Figure 5.6). Theoretical intrinsic availability represents the steady-state probability that a system is functioning under ideal maintenance conditions. In the plotted figure, as TTR increases, the availability of all machines gradually decreases and eventually stabilizes near their intrinsic availability levels. Machine M5, despite having a relatively high MTBF, exhibits the lowest availability due to a significantly higher MTTR, demonstrating the impact of poor maintainability. In contrast, machines M4, M8 and M10 maintain relatively high availability values (above 0.985), which indicates better repair responsiveness. Notably, M4, although experiencing more frequent failures, retains a good level of availability due to its fast and efficient maintenance, consistent with its high maintainability. These observations underscore the necessity of considering both MTBF and MTTR in maintenance planning, rather than relying solely on failure frequency.

TTR (h)	D(t)			
	M4	M5	M8	M10
10	0.9906	0.9926	0.9922	0.9918
40	0.9866	0.9823	0.9865	0.9851
70	0.9865	0.9793	0.986	0.9844
100	0.9865	0.9784	0.9859	0.9843
150	0.9865	0.9781	0.9859	0.9843
180	0.9865	0.978	0.9859	0.9843
200	0.9865	0.978	0.9859	0.9843
230	0.9865	0.978	0.9859	0.9843
270	0.9865	0.978	0.9859	0.9843
300	0.9865	0.978	0.9859	0.9843

Table 5.13. *Instantaneous availability calculation for hemodialysis machines*

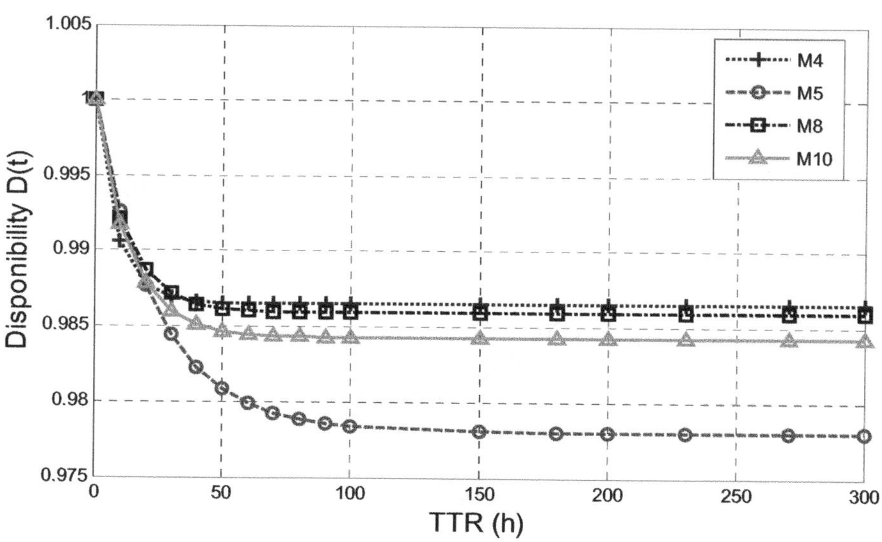

Figure 5.6. *Instantaneous availability for hemodialysis machines. For a color version of this figure, see www.iste.co.uk/bouchoucha/optimizing.zip*

5.8. The systematic inspection period

Continuous monitoring of failure data for dialysis machines is essential. This allows for the re-estimation of reliability parameters and the evaluation of the maintenance policy for the machine group. Such an approach can help reduce defects and repair times while improving overall system efficiency.

In practice, the main goal is to ensure maximum operational safety and reliability. Based on the desired reliability level, the systematic inspection interval (Tsys) for the dialysis machines under study can be determined using the reliability expression provided in equation [5.19] as follows:

$$Tsys = \eta \left(Log \frac{1}{R(t)} \right)^{\frac{1}{\beta}} \qquad [5.19]$$

The calculations carried out made it possible to determine the systematic inspection period of the hemodialysis machines. Figure 5.7 illustrates the systematic inspection period as a function of the reliability. The systematic inspection period is inversely proportional to the reliability R(t). In other words, to ensure effective reliability, machines must be checked more frequently. This study helps the maintenance department to set up a more precise and effective intervention plan.

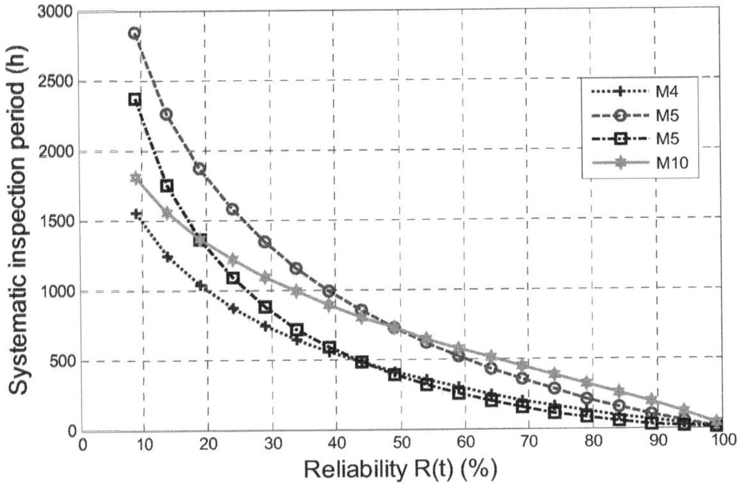

Figure 5.7. *Systematic inspection period for hemodialysis machines. For a color version of this figure, see www.iste.co.uk/bouchoucha/optimizing.zip*

Figure 5.7 illustrates the relationship between the required systematic inspection period and the desired reliability level for four different machines (M4, M5, M8 and M10). As expected, the inspection interval decreases sharply as the target reliability increases. This inverse relationship emphasizes that achieving high levels of reliability necessitates more frequent inspections. Machine M5, characterized by a high failure rate, requires the shortest inspection intervals to maintain a given reliability, while machine M10 allows for longer intervals due to its more stable performance. These findings are critical for establishing an effective maintenance plan, which defines the optimal inspection frequencies and intervention strategies based on the reliability behavior of each asset. A well-designed maintenance plan seeks to balance reliability goals with operational constraints by tailoring the inspection schedule to the failure characteristics of each machine. This ensures not only improved system availability but also cost-effective use of maintenance resources.

5.9. Development of a maintenance plan for hemodialysis machines

The principal aim of this book is to elaborate a maintenance plan for dialysis machines, taking into account the previous studies. This plan focuses on eliminating the 30% of the causes that caused 70% of the failures while maintaining a sufficient level of reliability, which means a systematic inspection period that must be calculated for each dialysis machine.

Table 5.14 represents a classification of the number of failures and downtime according to the causes of the failures. This classification makes it possible to determine the MTTR for each cause detected on all dialysis machines. According to the data collected in Table 5.14, it should be noted that the causes of failure that have a high MTTR, such as conductivity, acid pump and flow problems, require careful preparation of spare parts, appropriate tools and a qualified technician to intervene immediately and at an inferior cost. The high duration of the downtime observed for these causes is not directly linked to the maintenance department only but also to other external factors related to the general management of the company. Table 5.15 represents preventive maintenance tasks of dialysis devices to ensure the reliability and availability of the machines with the lowest cost. The execution of this plan requires the presence of qualified technicians as well as the necessary spare parts.

Failure / Machine	M4		M5		M8		M10		Total failure	Total TTR	MTTR
	TTR	N	TTR	N	TTR	N	TTR	N			
Flow parameter calibration problem	66.5	3	3.5	1	14	3	70	2	154	9	**17.1**
Pipe problem	3.5	1	0	0	0	0	0	0	3.1	1	3.1
Filter problem	3.5	1	10.5	1	3.5	1	0	0	17.5	3	5.83
Acid pump problem	10.5	2	0	0	87.5	2	3.5	1	101.5	5	**20.3**
Bicarbonate pump problem	0	0	0	0	24.5	3	7	1	31.5	4	7.78
Conductivity problem	7	2	119	2	3.5	1	3.5	1	133	6	**22.1**
Venous pressure nozzle problem	10.5	3	7	1	3.5	1	17.5	1	38.5	6	6.41
Water leak problem	0	0	7	1	0	0	7	2	14	3	4.66

Table 5.14. *TTR and failure number according to the failure causes for hemodialysis machines*

Machine	Tasks	Frequency
M4	1) Pressure/temperature sensor verification	Weekly
	2) Internal circuit cleaning	Bi-monthly
	3) Electrode calibration	Quarterly
M5	1) Flow and pressure control	Bi-monthly
	2) Filter cartridge replacement	Monthly
	3) Anti-scale cleaning	Monthly
	4) Visual inspection of seals and joints	Quarterly
M8	1) Vibration analysis of motor	Monthly
	2) Monitoring tube wear	Quarterly
	3) Load test for flow rate	Semi-annually
M10	1) Dialysate conductivity check	Weekly
	2) Early anomaly detection	Monthly
	3) Automatic calibration	Quarterly
	4) Dialysate pump maintenance	Quarterly

Table 5.15. *Preventive maintenance tasks of dialysis devices*

Detailed justification of each task

M4 – Dialysis machine (high criticality)

1) Sensor verification *(weekly)*
Predictive monitoring of pressure and temperature sensors ensures they remain within safe thresholds. Early deviation indicates possible clogging or component fatigue.
– *Justified by* real-time data analysis and sensor drift risk.

2) Internal circuit cleaning *(bi-monthly)*
Salt deposits and microbial biofilms can form in fluidic pathways, reducing flow efficiency or causing blockages.
– *Prevents* flow anomalies and contamination.

3) Electrode calibration *(quarterly)*
Accurate electrolyte detection is essential for patient safety. Over time, electrodes may degrade or drift.
– *Prevents* misdosing and alarms related to conductivity.

M5 – Water treatment (osmosis unit)

1) Flow/pressure control *(bi-monthly)*
It ensures that flow remains within expected ranges. Abnormal drops can indicate fouling or pump issues.
– *Protects* downstream dialysis machines.

2) Filter cartridge replacement *(monthly)*
It prevents clogging and drop in filtration efficiency. Cartridges have a known service life.
– *Classical preventive action* based on usage cycles.

3) Anti-scale cleaning *(monthly)*
Mineral deposits form even with pre-filtration. Regular descaling extends system life.
– *Recommended* in hard water environments.

4) Seal inspection *(quarterly)*
Leaks often originate from degraded joints. Visual inspections help avoid water damage and contamination.
– *Essential for* long-term integrity.

M8 – Peristaltic pump (medium criticality)

1) Motor vibration analysis *(monthly)*
Imbalance or early mechanical wear causes specific vibration signatures. Vibration sensors detect anomalies.
– *Prevents* motor failure and unplanned downtime.

2) Tube wear monitoring *(quarterly)*
Tubes degrade over time due to flex cycles. Using Weibull distributions helps determine optimal replacement intervals.
– *Predictive task* based on failure probability.

3) Flow load test *(semi-annually)*
It confirms actual output matches design specs under simulated load. Detects performance degradation not visible at idle.
– *Useful for* verifying system capacity and reliability.

M10 – Dialysate generator (very high criticality)

1) Conductivity check *(weekly)*
It ensures dialysate concentration is safe. Deviations could harm patients.
– *Critical safety check* should not be delayed.

2) Early anomaly detection (Bayesian) *(monthly)*
Uses networked models to correlate small parameter changes and estimate failure probabilities.
– *Improves accuracy* of preventive actions.

3) Automatic calibration *(quarterly)*
It keeps sensors and flow controllers aligned with reference values. Important for safe operation.
– *Reduces human error* and maintains precision.

4) Pump maintenance *(quarterly)*
It includes lubrication, seal check and wear assessment. Pumps are key moving parts subject to fatigue.
– *Protects* against catastrophic failure.

5.10. Conclusion

In this chapter, a numerical study of the dialysis system failure statistics was presented. The working principle and different parts of the dialysis machine were explored and discussed. The statistical count of all failures detected on each dialysis machine was determined and included in the system analysis. The Pareto analysis method was introduced to determine the priority of the maintenance department interventions by identifying the most serious failures that affect the operational safety of the machines. The Weibull distribution was used to study the reliability of the hemodialysis system. The Weibull parameters were identified, and the efficiency of the chosen distribution was validated through the Kolmogorov–Smirnov test. Among the contributions of this work, the maintainability and availability of the dialysis system were evaluated. These can be effective tools for the assessment of the current situation and for the prediction of the reliability of the upgrade of the operational management of the dialysis system. In addition, the question of the systematic inspection period was treated. At the end of this chapter, a plan of preventive maintenance was developed with the objective of planning the different types of interventions to guarantee maximum reliability and availability of the dialysis machines.

For future studies, it is essential to implement a continuous monitoring system for the control and diagnosis of risks in dialysis machines. Such a system could help reduce failures and repair times while enhancing both the reliability and availability of the equipment. This research can be extended in several directions, including the analysis of operational safety for dialysis systems in stochastic environments under uncertain conditions. Traditional reliability methods are often insufficient to handle these uncertainties. Therefore, the maintenance department should explore alternative approaches for monitoring device reliability in complex and dynamic environments. In particular, the application of artificial intelligence (AI) techniques to provide continuous supervision of dialysis machines represents a promising area for detailed and rigorous investigation.

6

Effect of the Stochastic Weibull Parameters on the Reliability of the Dialysis Machine

6.1. Introduction

Dialysis machines operate in environments characterized by significant uncertainty and multiple sources of disturbances, making it essential to account for these uncertainties throughout their entire lifecycle. This chapter presents a study on the stochastic reliability of dialysis machines using the two-parameter Weibull distribution. The shape and scale parameters were estimated using both analytical and graphical methods based on the recorded failure histories of the devices. One of the main objectives of this work was to formulate these analytical methods and implement them through numerical simulations.

The uncertainties in the machines' operational environment arise from variability in physical and geometric parameters, fluctuations in load conditions, boundary stresses and the simplifications or assumptions made during the modeling process. These uncertainties are captured in the Weibull parameters, and their impact on device reliability was systematically analyzed. The study particularly focused on the effect of uncertainties in the Weibull parameters, treating the involved parameters as Gaussian variables and calculating their means and standard deviations for various

For a color version of all the figures in this chapter, see www.iste.co.uk/bouchoucha/optimizing.zip.

configurations of the dialysis machines. The random failure rates and reliability functions were evaluated and discussed. Furthermore, the concept of a random systematic inspection interval was investigated to design an effective preventive maintenance program.

Modern medical devices and materials have become highly sophisticated, incorporating advanced technologies and often operating under challenging conditions. Healthcare facilities must ensure that their essential medical equipment is accurate, safe, reliable, available and efficient. Surgical devices, in particular, are critical for maintaining a high standard of care in medical institutions. In developed hospitals, the number of different types of medical devices exceeds 5,000 (Dhillon 2000). Given the performance expectations in contemporary healthcare and the increasing complexity of medical equipment, many countries place significant emphasis on enhancing the quality and reliability of devices used in hospitals. According to the World Health Organization (WHO), approximately 50% of medical equipment in developing countries is not functioning properly, primarily due to inadequate maintenance, aging and deterioration of equipment, and diminished service quality (World Health Organization, Regional Committee for the Eastern Mediterranean 2006).

Failures of medical devices can disrupt healthcare operations, pose serious health risks to patients and even negatively impact the environment (Zamzam et al. 2021). A failure is defined as a partial or complete loss of a component's essential properties, resulting in a significant reduction or total loss of its operational capability. Such failures may arise from design flaws, manufacturing defects, improper installation, misuse, or inadequate maintenance practices (Arunraj and Maiti 2007; Brissaud et al. 2010).

All production systems are subject to aging and wear over time, affecting their functional components (Valdez-Flores and Feldman 1989). These physical phenomena can lead to failures that substantially increase operating costs and compromise operational safety. The literature presents various classifications of failures (Deloux et al. 2008; Rausand et al. 2020) and investigates the effects of fatigue and aging in mechanical systems on availability, maintainability and reliability (Alfares 1999; Gorjian et al. 2010).

Equipment degradation can result from multiple factors, including operating conditions, vibrations, fatigue, wear and applied stresses. Modeling time-dependent failure rates is therefore essential for monitoring and assessing the risks associated with sensitive medical equipment. Gorjian et al. (2010) analyzed the diverse and often complex mechanisms of mechanical component degradation, including cracking, wear, creep and fatigue. These degradation modes are influenced by numerous parameters, such as material properties, geometric and dimensional characteristics, operating conditions and external stresses (Aven et al. 2006).

Influence factors are internal or external parameters that affect the reliability of a machine. Brissaud et al. (2007) proposed a classification of these factors according to the life stage of the system under consideration, including categories such as design, manufacturing, system installation, operational and maintenance factors. Organizational and human factors, often having broader system-wide effects, can also be added to this list (Cui and Li 2019). These factors are part of a multidisciplinary domain focused on enhancing safety, improving performance and increasing user satisfaction. They encompass areas such as operational procedures, training and competency, safety-critical communication, health and safety culture, maintenance, inspection and testing errors (Alvarenga et al. 2014).

In the context of medical equipment, Dhillon (2006) emphasized that maintenance is essential to preserve devices in optimal working condition, ensure reliability or restore them to a defined operational state. Mathematical maintenance approaches can further improve equipment reliability through various procedures and techniques. Bahreini et al. (2018) aimed to identify the factors influencing medical equipment maintenance management. A well-designed maintenance strategy reduces downtime and associated costs while enhancing productivity, profitability and competitiveness by ensuring efficient equipment operation (Al-Najjar and Alsyouf 2004).

Furthermore, Jiménez et al. (2023) proposed an ontology-based model for selecting and assessing maintenance strategies, while Sielaffa and Lucke (2021) and Liao et al. (2021) developed approaches for integrated maintenance strategy selection that account for machine failures and the context of other machines within a value-adding network.

In modern engineering applications, it is essential to carefully select the probability distribution that best represents a dataset, especially when dealing with uncertain parameters (O'Connor and Kleyner 2012). The Weibull distribution is among the most widely used probability density functions in industry. Lyonnet (1991) highlighted that the Weibull model is particularly well-suited for reliability analysis of mechanical components. Its main advantage lies in its ability to provide meaningful insights even when failure history data are limited. For datasets of small size, graphical methods are recommended to ensure the accuracy and precision of the probabilistic analysis (Zhang and Xie 2007).

The Weibull distribution is flexible, allowing it to model various failure behaviors and to approximate other statistical distributions commonly employed in engineering applications (Lihou and Spence 1988). It is extensively used for life data analysis across a range of fields, including civil aviation, aerospace, automotive, electronics and materials engineering (Luko 1999; Fidanoglu et al. 2017). Additionally, this statistical approach can effectively assess the reliability of semiconductors, ball bearings, spot welds, engines and even biological systems (Rausand et al. 2020).

Reliability analysis using a purely deterministic approach is often insufficient to capture the behavior of random data in a stochastic environment. In such contexts, uncertain reliability parameters are typically represented by their most unfavorable values. Therefore, predictive modeling of system behavior should rely on probabilistic approaches (Fenina et al. 2024a). The uncertainties and errors observed in the results are associated with variability in physical and geometric parameters, boundary stress conditions, poorly characterized load fluctuations as well as simplifying assumptions and physical laws adopted in the modeling process (Tebbi et al. 2005). Several studies (Gouno and Guérineau 2015) have shown that these influencing factors primarily affect the shape and scale parameters of the Weibull distribution. This work accounts for uncertainties arising from the chosen methods, modeling approximations and influential variables in the Weibull parameters, and their impact on the reliability of the device was thoroughly analyzed.

This chapter presents a study on the stochastic reliability of dialysis machines using the two-parameter Weibull distribution. In a stochastic environment, the reliability of the device is influenced by random effects. These uncertainties, which arise from external factors, are diverse and challenging to model accurately due to their complex and unpredictable nature. In this work, an approach is proposed that incorporates the impact of influencing factors on reliability. The methodology involves introducing perturbations to the shape and scale parameters of the Weibull distribution and analyzing their effect on key reliability indicators. The Weibull parameters are estimated using analytical methods based on the recorded failure histories of the devices. These parameters are treated as random variables following a Gaussian distribution, and their means and standard deviations are calculated. Numerical simulations are then used to present the statistical characteristics of the dialysis machine's random reliability.

6.2. Estimation of the stochastic Weibull parameters

The reliability of dialysis devices plays a crucial role in guaranteeing patient safety as well as the continuity and effectiveness of treatment. According to Azar (2009), a significant relationship exists between patient outcomes and the maintenance of dialysis machines, highlighting that insufficient maintenance can lead to serious complications and inappropriate dialysis sessions. Consequently, establishing a straightforward and effective maintenance strategy is essential to ensure both the safety and efficiency of dialysis procedures. A comprehensive review of the maintenance and inspection of medical devices was presented by Jamshidi et al. (2014). In addition, Alnazier et al. (2013) investigated the electrical and mechanical failures occurring in dialysis equipment, analyzing their root causes and proposing preventive measures. Tsarouhas (2019) further explored the reliability and maintainability of dialysis machines by applying the Weibull distribution.

In this section, the stochastic reliability of 10 dialysis machines (Mi, i = 1:10) is examined. The failure records of these devices were gathered over a 9-year period, from 2013 to 2022. A synthesis of this failure history is presented in Tables 6.1(a–c). Each machine operates on a schedule of 3.5 h per day, 6 days a week, corresponding to 288 operational days annually. To illustrate, one machine commissioned on January 1, 2013 experienced its first failure on June 5, 2015. This corresponds to 700 days of functioning, representing an uptime of 2,450 h. The estimation of the Weibull parameters, using both graphical and analytical approaches, is reported in Table 6.2. These shape and scale parameters were derived directly from the collected failure data, applying the methods discussed in the previous chapters.

The operating environment of dialysis machines is inherently uncertain. Similar forms of uncertainty are also encountered in other domains such as healthcare and disease management. These uncertainties arise not only from methodological aspects in the formulation stages but also from the approximations introduced in mathematical modeling and development processes (Bouchoucha et al. 2017). Furthermore, the presence of missing values within datasets can significantly influence the accuracy of the outcomes. External influencing factors constitute a major source of disturbance and, as highlighted by Hajji (2019), can be categorized into several groups. These include individual-related aspects such as technical skills, competence, training, motivation, stress or the physical and psychological state of the operator. Other categories concern team-related factors, including communication and supervision; procedural factors, such as scheduling, dispatching and availability; and environmental factors, which encompass materials, infrastructure, equipment, supplies and overall working conditions.

In traditional reliability analysis, the shape and scale parameters are generally represented by deterministic values corresponding to the least favorable conditions (Fenina et al. 2024b). To overcome this limitation, the two Weibull parameters are instead modeled as random variables that follow a Gaussian distribution (Appendix 3), as expressed in equation [6.1]:

$$\tilde{\beta} = \beta + \sigma_\beta \varepsilon \text{ and } \tilde{\eta} = \eta + \sigma_\eta \varepsilon \qquad [6.1]$$

Here, β and η represent the mean values of the random shape and scale parameters, respectively. The terms σ_β and σ_η denote the standard deviations associated with these parameters, while ε corresponds to a reduced centered Gaussian variable (Appendix 3). The mean value of the parameters is determined according to equation [6.2]:

$$\beta = (1/N)\sum_{i=1}^{N} \beta_i \text{ and } \eta = (1/N)\sum_{i=1}^{N} \eta_i \qquad [6.2]$$

In this context, β_i and η_i correspond to the shape and scale parameters, respectively, for the ith draw, while N represents the total number of draws.

The standard deviations of the shape and scale parameters are calculated using equation [6.3]:

$$\sigma_\beta = \frac{1}{N^2}\sqrt{\sum_{i=1}^{N}(\beta_i - \beta)^2} \text{ and } \sigma_\eta = \frac{1}{N^2}\sqrt{\sum_{i=1}^{N}(\eta_i - \eta)^2} \qquad [6.3]$$

Table 6.3 presents the mean values and standard deviations of the shape and scale parameters estimated for each dialysis device (Fenina et al. 2023). These calculations are performed following equations [6.2] and [6.3]. The N draws correspond to the four methods used to determine β and η, namely, the graphical method (GM), least squares method (LSM), maximum likelihood method (MLM) and method of moments (MOM). Based on the numerical results in Table 6.3, the dialysis machines can be classified according to their operational life phase (Fenina et al. 2023). Specifically, devices M5 and M10, with shape parameters $\beta > 1$, are in the aging phase. Machines M1, M2, and M6, with $\beta \approx 1$, operate in the useful life phase. Devices M3, M4, M7, M8, and M9, with $\beta < 1$, are in the early-life phase.

This case study highlights the behavior of Weibull parameters under stochastic conditions. The use of graphical and analytical methods to estimate the shape and scale parameters introduces potential sources of error due to modeling assumptions and mathematical approximations. These uncertainties allow the computation of the mean and standard deviation of the Weibull parameters. Accordingly, this study focuses on evaluating the impact of uncertainties in the shape and scale parameters on the reliability of the dialysis machines.

No.	M1		M2		M3	
	Failure date	**TBF (h)**	**Failure date**	**TBF (h)**	**Failure date**	**TBF (h)**
1	05/06/2015	2,450	05/06/15	2,450	29/01/14	1,085
2	09/11/2015	427	20/11/17	2,481.5	04/01/16	1,904
3	20/11/2017	2,047	23/01/19	1,155	19/10/16	609
4	27/07/2019	1,673	25/03/19	169	24/10/16	10.5
5	06/03/2020	595	06/03/20	1,039.5	29/05/17	605.5
6	05/09/2021	1,372			06/03/20	2,792
7	01/11/2021	147			12/11/20	392

Table 6.1. *(a) The failure history (TBF) of dialysis machines (M1 to M3)*

No.	M4		M5		M6	
	Failure date	TBF (h)	Failure date	TBF (h)	Failure date	TBF (h)
1	17/04/15	2,313.5	05/06/15	2,450	05/06/2015	2,450
2	05/06/15	119	15/05/17	1,949.5	29/05/2017	2,040
3	29/05/17	1,988	20/11/17	5,14.5	07/08/2018	192.5
4	06/12/17	525	06/03/20	2,268	28/02/2018	549.5
5	06/03/20	3,276	18/09/20	385	03/08/2018	311.5
6	24/06/20	154	16/09/21	990.5	06/03/2020	584.5
7					03/08/2020	280

Table 6.1. *(b) The failure history (TBF) of dialysis machines (M4 to M6)*

No.	M7		M8		M9	
	Failure date	TBF (h)	Failure date	TBF (h)	Failure date	TBF (h)
1	04/01/16	3,034.5	13/03/15	2,226	22/01/13	73.5
2	29/05/17	1,389	04/01/16	766.5	15/07/13	448
3	12/06/17	42	13/01/16	3.5	01/03/14	717.5
4	20/11/17	441	22/07/13	535.5	08/10/14	578
5	15/10/18	871.5	19/09/16	140	05/06/15	3,650.5
6	11/09/19	903	02/08/17	878.5	29/05/17	192.5
7	06/03/20	486.5	08/01/18	434	07/08/17	553
8	29/03/21	896	05/02/18	66.5	28/02/18	73.5
9	22/07/21	304.5	28/02/18	63	07/05/18	213.5
10			10/08/18	336	03/08/18	28
11			06/03/20	1,575	19/11/18	294
12			08/07/20	203	18/09/19	829.5
13			24/02/21	633.5	06/03/20	462
14			05/11/21	703.5	02/12/20	560

Table 6.1. *(c) The failure history (TBF) of dialysis machines (M7 to M9)*

No.	M10	
	Failure date	**TBF (h)**
1	29/05/13	413
2	12/07/13	101.5
3	01/09/14	1,130
4	05/06/15	763
5	09/11/15	427
6	07/08/17	1,750
7	31/10/18	1,228
8	09/08/19	780.5
9	06/03/20	560
10	29/07/20	255.5
11	07/01/22	1,456
12		
13		
14		

Table 6.1. *(d) The failure history (TBF) of dialysis machine M10*

	LSM		MLM		MOM		GM	
	β	η	β	η	β	η	β	η
M1	0.928	1,363	1.078	1,504	1.179	1,583	0.963	1,358
M2	0.752	1,957	1.058	1,814	1.184	2,028	0.866	1,672
M3	0.388	1,253	0.723	1,472	0.658	1,356	0.421	1,214
M4	0.625	1,342	0.859	1,665	0.774	1,552	0.586	1,338
M5	1.232	1,642	1.318	1,764	1.405	1,802	1.161	1,524
M6	1.027	904	1.052	1,036	1.061	1,125	0.908	879
M7	0.844	934	0.892	1,148	0.954	1,176	0.862	970
M8	0.625	572	0.694	728	0.762	791	0.607	473
M9	0.785	486	1.024	814	0.998	745	0.842	573
M10	1.162	887	1.387	952	1.638	1,050	1.105	823

Table 6.2. *Estimation of the random Weibull parameters of dialysis machines*

	Mean β	Mean η	σ_β	σ_η	Life phase
M1	1.037	1,452	0.099	95.684	Useful
M2	0.965	1,868	0.167	136.8	Useful
M3	0.548	1,324	0.206	208.26	Early
M4	0.711	1,557	0.11	162.7	Early
M5	1.279	1,683	0.091	109.18	Aging
M6	1.012	986	0.062	99.992	Useful
M7	0.888	1,057	0.042	106.23	Early
M8	0.672	641	0.061	125.55	Early
M9	0.912	654	0.101	131	Early
M10	1.323	928	0.210	83.913	Aging

Table 6.3. *The mean and the standard deviation of the shape and scale parameters*

6.3. Random reliability analysis

6.3.1. *The random failure rate estimation*

The random failure rate can be formulated as a random variable according the Gaussian distribution (Fenina et al. 2023) as follows (equation [6.4]):

$$\tilde{\lambda}(t) = \lambda(t) + \sigma_\lambda(t)\varepsilon \qquad [6.4]$$

Here, λ(t) is the mean of the failure rate and $\sigma_{\lambda(t)}$ is its standard deviation.

According to the two-parameter Weibull distribution, the random failure rate can be expressed through the shape and scale parameter as given in equation [6.5]:

$$\tilde{\lambda}(t) = (\tilde{\beta}/\tilde{\eta})(t/\tilde{\eta})^{\tilde{\beta}-1} \qquad [6.5]$$

To extract the mean and the standard deviation of the failure rate, the logarithmic failure rate can be expressed according to equation [6.6]:

$$ln\left(\tilde{\lambda}(t)\right) = ln(\tilde{\beta}) - \tilde{\beta}\, ln(\tilde{\eta}) + (\tilde{\beta} - 1)\, ln(t) \qquad [6.6]$$

To linearize the random failure rate formulated in equation [6.6], the first-order (Ichchou et al. 2011) Taylor series expansion is used (Appendix 4) to extract the random quantity of log(λ(t)), log(β) and log(η) in the vicinity of their mean, respectively, according to equations [6.7]–[6.9]:

$$ln\left(\tilde{\lambda}(t)\right) = ln(\lambda(t)) + (\sigma_\lambda(t)/\lambda(t))\,\varepsilon \quad [6.7]$$

$$ln(\tilde{\beta}) = ln(\beta) + (\sigma_\beta/\beta)\,\varepsilon \quad [6.8]$$

$$ln(\tilde{\eta}) = ln(\eta) + (\sigma_\eta/\eta)\,\varepsilon \quad [6.9]$$

Introducing equations [6.7]–[6.9] in equation [6.6], we can write equation [6.10]:

$$ln\,\lambda + (\sigma_\lambda/\lambda)\,\varepsilon = [ln(\beta) - \beta\, ln(\eta) + (\beta - 1)\, ln(t)]$$

$$+\left[(\sigma_\beta/\beta) - (\beta/\eta)\sigma_\eta - \sigma_\beta\, ln(\eta) + \sigma_\beta\, ln(t)\right]\varepsilon \quad [6.10]$$

By identifying the various terms in equation [6.10], it becomes possible to determine the mean and standard deviation of the random failure rate, as expressed in equations [6.11] and [6.12], respectively:

$$ln(\lambda(t)) = ln(\beta) - \beta\, ln(\eta) + (\beta - 1)\, ln(t) \quad [6.11]$$

$$\sigma_\lambda(t)/\lambda(t) = \sigma_\beta/\beta - (\beta/\eta)\sigma_\eta - \sigma_\beta\, ln(\eta) + \sigma_\beta\, ln(t) \quad [6.12]$$

The mean and standard deviation of the random failure rate are calculated using equations [6.13] and [6.14], respectively:

$$\lambda(t) = (\beta/\eta)(t/\eta)^{\beta-1} \quad [6.13]$$

$$\sigma_\lambda(t) = \lambda(t)\left[\sigma_\beta/\beta - (\beta/\eta)\sigma_\eta - \sigma_\beta\, ln(\eta) + \sigma_\beta\, ln(t)\right] \quad [6.14]$$

Figure 6.1 presents the mean (a) and standard deviation (b) of the failure rate for dialysis machines in the aging phase (M5 and M10), with classification according to life phase (Fenina et al. 2024a). For machines M5 and M10, which are in the aging phase ($\beta > 1$), the mean failure rate λ(t) exhibits a continuous increase over time, indicating the need for constant monitoring to prevent potential failures. Figure 6.1(b) shows the standard deviation of the failure rate for these aging machines, highlighting the propagation of uncertainties introduced in the shape and scale parameters. The results indicate that these uncertainties are amplified over time, significantly disturbing the failure rates and emphasizing the importance of accurate determination of the Weibull parameters.

Figure 6.2 illustrates the mean (a) and standard deviation (b) of the failure rate for machines in the useful life phase (M1, M2 and M6). In this phase, the mean failure rate remains constant over time ($\beta \approx 1$). However, Figure 6.2(b) shows that the standard deviation of the failure rate increases significantly for M1, M2 and M6, indicating that uncertainties accumulate and amplify even when the mean failure rate is stable. These findings demonstrate that uncertainties in the Weibull parameters impact the failure rates of machines in their useful life, underscoring the necessity for the maintenance department to ensure precise estimation of the shape and scale parameters in any stochastic environment, regardless of the machine's life phase.

Machines M3, M4, M7, M8 and M9, which are in the early-life phase ($\beta < 1$), exhibit a decreasing mean failure rate as shown in Figure 6.3(a). The standard deviation of the failure rate for these early-life machines is depicted in Figure 6.3(b). The numerical results reveal the significant influence of uncertainties in the Weibull parameters, which can markedly affect the failure rate and, consequently, the operational safety of the machines across all life phases.

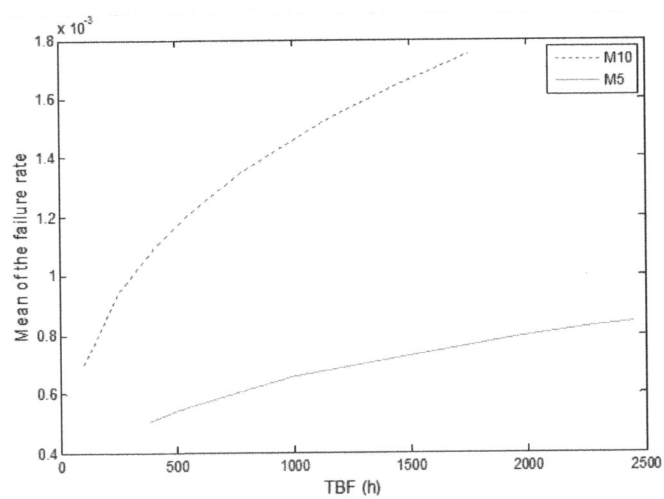

Figure 6.1. *(a) Mean of the failure rate of dialysis machines in aging life*

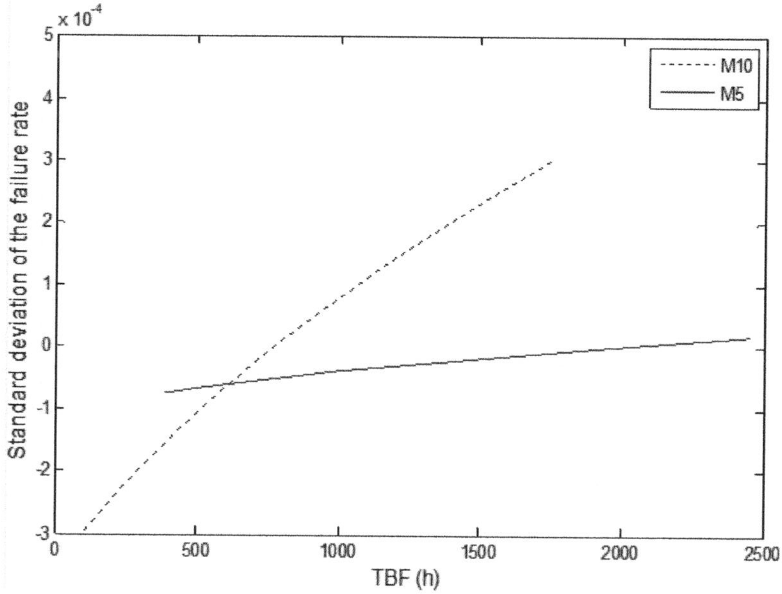

Figure 6.1. *(b) Standard deviation of the failure rate of dialysis machines in aging life*

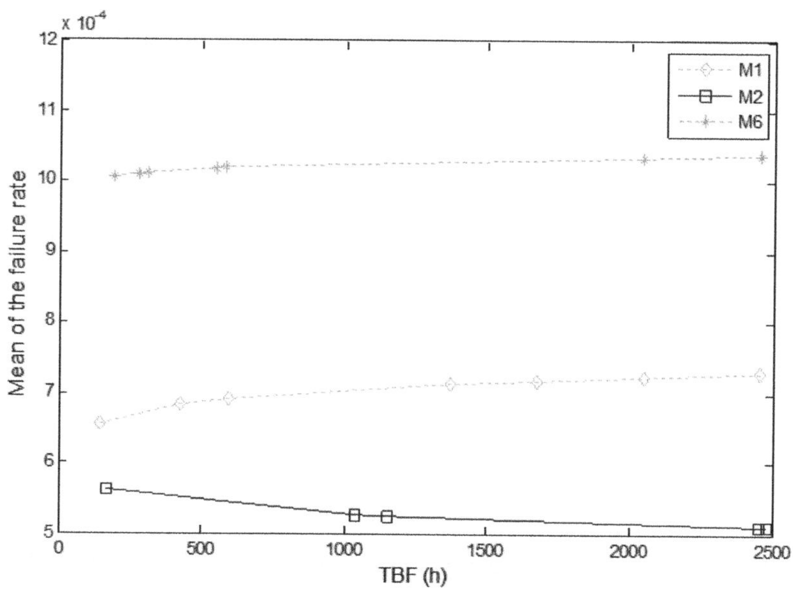

Figure 6.2. *(a) Mean of the failure rate of dialysis machines in useful life*

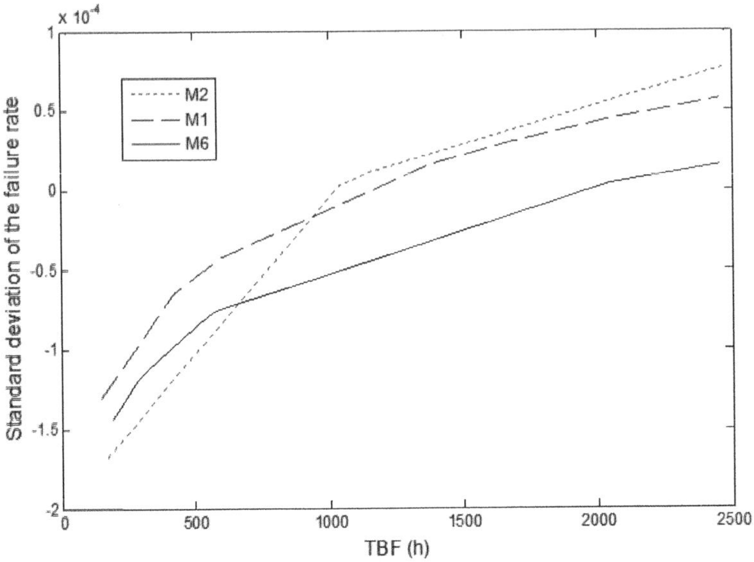

Figure 6.2. *(b) Standard deviation of the failure rate of dialysis machines in useful life*

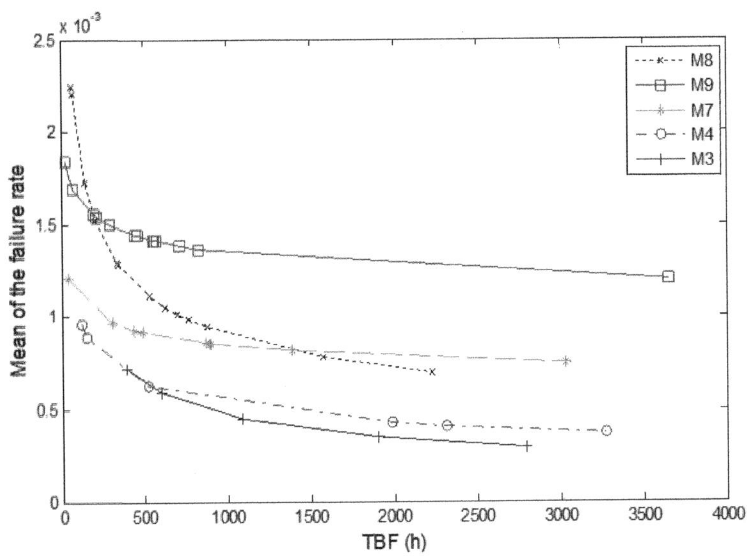

Figure 6.3. *(a) Mean of the failure rate of dialysis machines in early life*

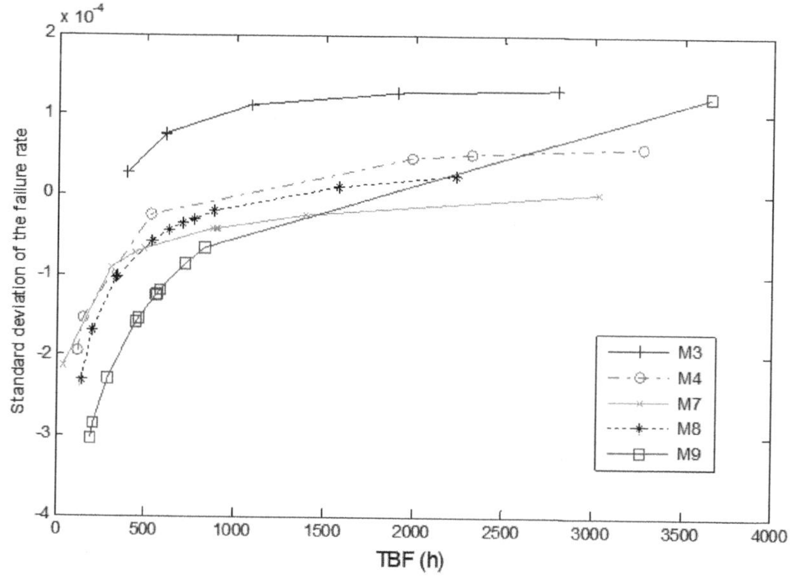

Figure 6.3. *(b) Standard deviation of the failure rate of dialysis machines in early life*

6.3.2. *The random reliability estimation*

The random reliability is formulated as a random variable according to the Gaussian distribution as given in equation [6.15]:

$$\tilde{R} = R(t) + \sigma_R(t)\varepsilon \qquad [6.15]$$

where R(t) is the mean of the failure rate and $\sigma_R(t)$ is its standard deviation.

Through the two-parameter Weibull distribution, the random reliability is given as a function of the scale and shape parameter as given in equation [6.16]:

$$\tilde{R}(t) = exp(-t/\tilde{\eta})^{\tilde{\beta}} \qquad [6.16]$$

The first-order Taylor series expansion of $\widetilde{R}(t)$ is given in equation [6.17] (Fenina et al. 2024b):

$$\tilde{R}(t) = exp(-t/\tilde{\eta})^{\tilde{\beta}} = 1 + (-t/\tilde{\eta})^{\tilde{\beta}} \qquad [6.17]$$

The logarithmic reliability is expressed in the following equation [6.18]:

$$Ln\big(\tilde{R}(t) - 1\big) = \tilde{\beta}(Ln(-t) - Ln(\tilde{\eta})) \quad [6.18]$$

To linearize the random logarithmic reliability equation, a first-order Taylor series expansion around the mean is employed (Appendix 4). The first and second terms of equation [6.18] are expanded as shown in equations [6.19] and [6.20], respectively.

$$Ln\big(\tilde{R}(t) - 1\big) = Ln(R(t) - 1) + (\sigma_R(t)/R(t))\varepsilon \quad [6.19]$$

$$\tilde{\beta}\big(Ln(-t) - Ln(\tilde{\eta})\big) = \beta Ln(-\, t/\tilde{\eta}) + \big(\sigma_\beta Ln(-\, t/\tilde{\eta}) - (\beta/\eta)\sigma_\eta\big)\varepsilon \quad [6.20]$$

By substituting the first and second terms of equation [6.18] with their respective expansions from equations [6.19] and [6.20], the equation can be rewritten as follows:

$$Ln(R(t) - 1) + (\sigma_R(t)/R(t))\varepsilon =$$

$$\beta Ln(-\, t/\tilde{\eta}) + \big(\sigma_\beta Ln(-\, t/\tilde{\eta}) - (\beta/\eta)\sigma_\eta\big) \quad [6.21]$$

By identifying the various terms in equation [6.21], it becomes possible to determine the mean and standard deviation of the random reliability, as expressed in equations [6.22] and [6.23], respectively.

$$Ln(R(t) - 1) = \beta Ln(-\, t/\tilde{\eta}) \quad [6.22]$$

$$\sigma_R(t)/R(t) = \sigma_\beta Ln(-\, t/\tilde{\eta}) - (\beta/\eta)\sigma_\eta \quad [6.23]$$

The mean and the standard deviation of the random reliability are formulated in equations [6.24] and [6.25], respectively:

$$R(t) = exp(-\, t/\eta)^\beta \quad [6.24]$$

$$\sigma_R(t) = R(t)\big(\sigma_\beta Ln(-\, t/\eta) - (\beta/\eta)\sigma_\eta\big) \quad [6.25]$$

The random cumulative density function is formulated according to equation [6.26] as a random variable following a Gaussian distribution:

$$\tilde{F}(t) = 1 - \tilde{R}(t) = F(t) + \sigma_F\varepsilon \quad [6.26]$$

where F(t) is the mean of the cumulative density function and $\sigma_{F(t)}$ is its standard deviation, presented by equation [6.27]:

$$F(t) = 1 - R(t) \text{ and } \sigma_F(t) = \sigma_R(t) \quad [6.27]$$

Figures 6.4(a and b), 6.5(a and b) and 6.6(a and b) display the mean and standard deviation of the reliability and cumulative density function for dialysis machines in the aging, useful and early-life phases. The simulations of the mean reliability for the machines are presented in Figures 6.4(a), 6.5(a) and 6.6(a). According to this study, the reliability R(t) of machines M5 and M10 requires immediate review, as their reliability declines significantly over time and demands continuous monitoring (Figure 6.4(a)). The mean reliability of M1, M2 and M6, depicted in Figure 6.5(a), shows a gradual decrease over time, indicating the importance of adhering to preventive maintenance to prolong the operational life of both the machines and their components. Figure 6.6(a) illustrates the mean reliability of five machines in the early-life phase, which are functioning correctly and need to complete a trial period to achieve optimal performance.

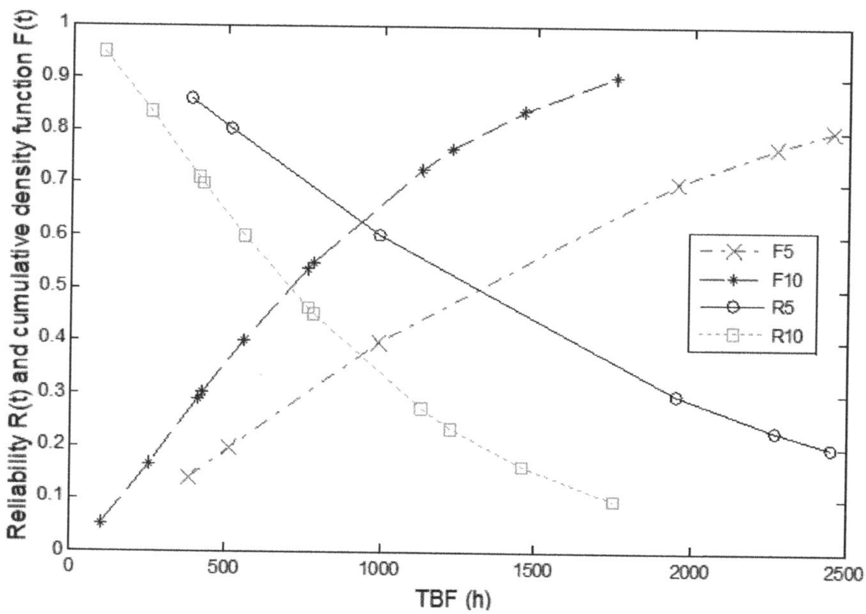

Figure 6.4. *(a) Mean of reliability and cumulative density function of machines in aging life*

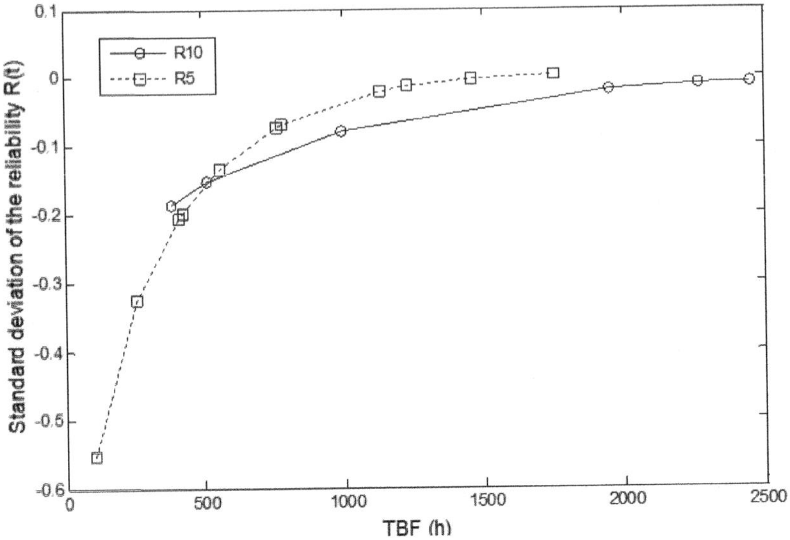

Figure 6.4. *(b) Standard deviation of reliability of machines in aging life*

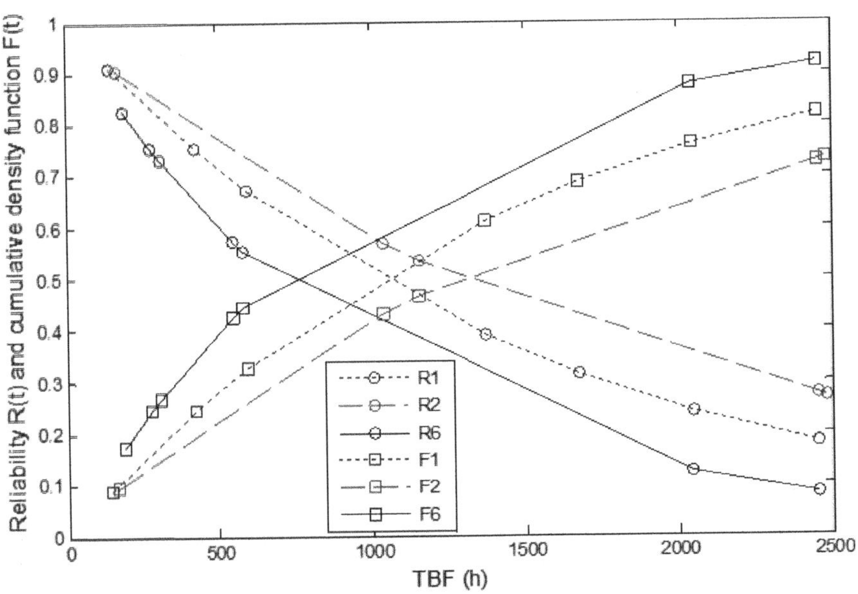

Figure 6.5. *(a) Mean of reliability and cumulative density function of machines in useful life*

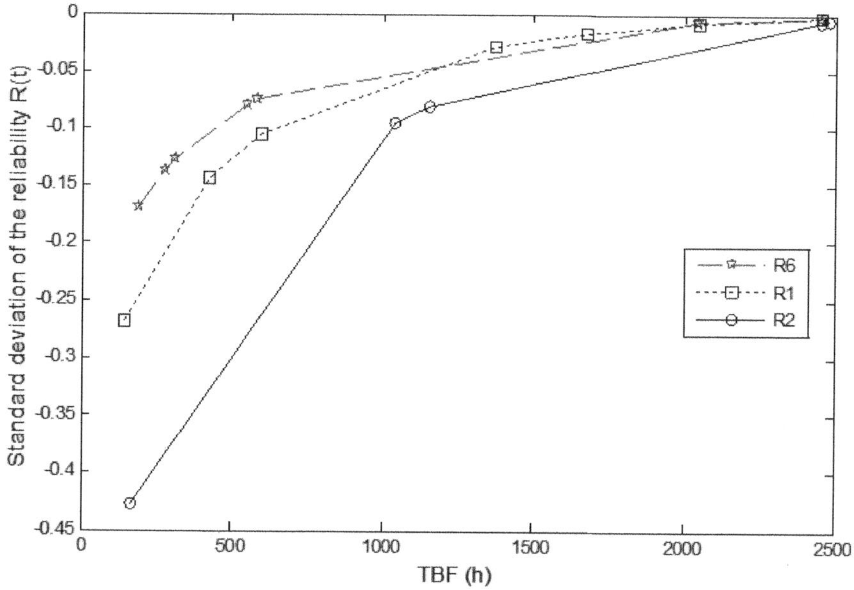

Figure 6.5. *(b) Standard deviation of reliability of machines in useful life*

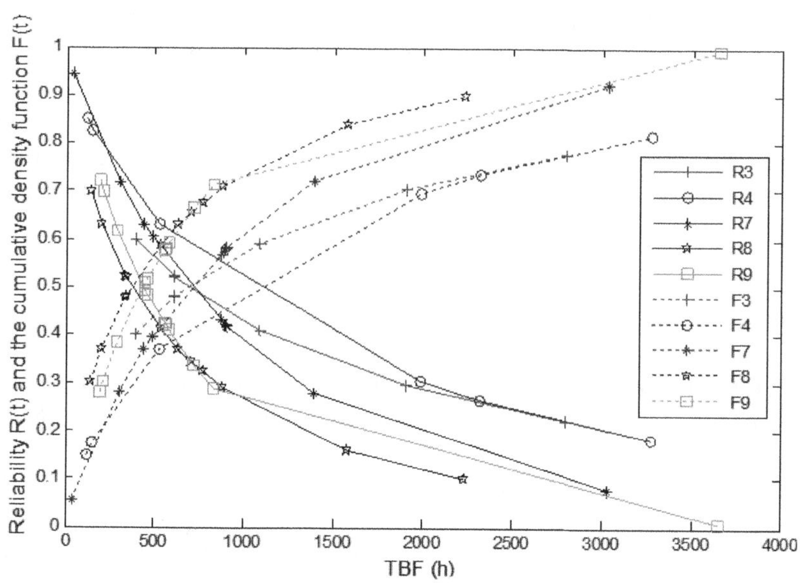

Figure 6.6. *(a) Mean of reliability and cumulative density function of machines in early life*

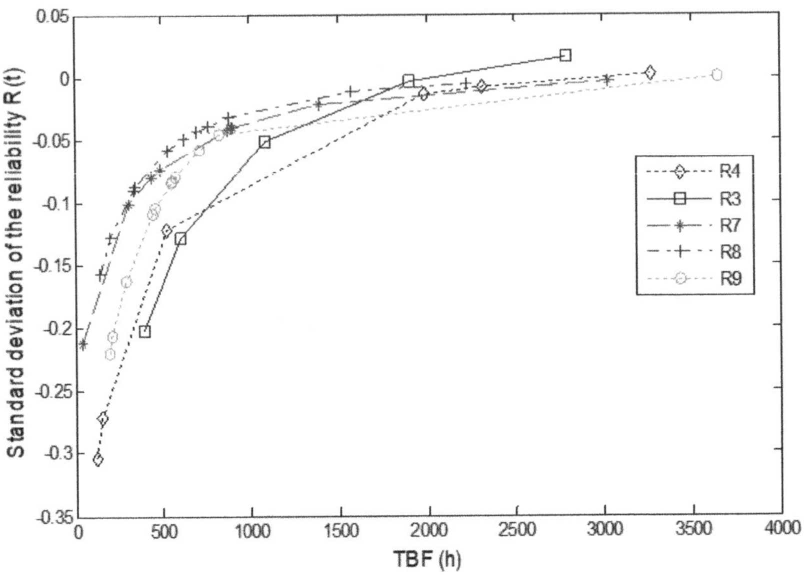

Figure 6.6. *(b) Standard deviation of reliability of machines in early life*

The simulations of the standard deviation of the reliability, shown in Figures 6.4(b), 6.5(b) and 6.6(b), reveal an increasing trend over time, indicating that the impact of uncertainties becomes progressively more significant. The reliability of the machines is strongly influenced by uncertainties in the Weibull parameters. The level of disturbance introduced in the shape and scale parameters is amplified, presenting a considerable risk to the safe and correct operation of the devices.

6.3.3. *The random probability density function estimation*

The random probability density function is formulated as a random variable following a Gaussian distribution presented in equation [6.28]:

$$\tilde{f}(t) = \tilde{\lambda}(t) \times \tilde{R}(t) = f(t) + \sigma_f(t)\varepsilon \qquad [6.28]$$

The development of the random variables permits us to write equation [6.29]:

$$f(t) + \sigma_f(t)\varepsilon = \lambda(t)R(t) + \big(R \times \sigma_\lambda(t) + \lambda \times \sigma_R(t)\big)\varepsilon \qquad [6.29]$$

The mean (equation [6.30]) and the standard deviation (equation [6.31]) of the random probability density function can be formulated by identifying the different terms in equation [6.29]:

$$f(t) = \lambda(t) \times R(t) \qquad [6.30]$$

$$\sigma_f(t) = R(t) \times \sigma_\lambda(t) + \lambda(t) \times \sigma_R(t) \qquad [6.31]$$

Figures 6.7(a and b), 6.8(a and b) and 6.9(a and b) depict the mean and standard deviation of the probability density function (PDF) for dialysis machines across the different life phases: aging (Figure 6.7), useful (Figure 6.8) and early-life (Figure 6.9). The mean PDF of the dialysis machines is illustrated in Figures 6.7(a), 6.8(a) and 6.9(a), highlighting the influence of the life phase on the distribution. The numerical simulations were carried out using equation [6.30] to compute the mean of the PDF and equation [6.31] to determine its standard deviation. Figures 6.7(b), 6.8(b) and 6.9(b) present the standard deviation of the PDF, showing that the stochastic environment progressively amplifies its variability over time.

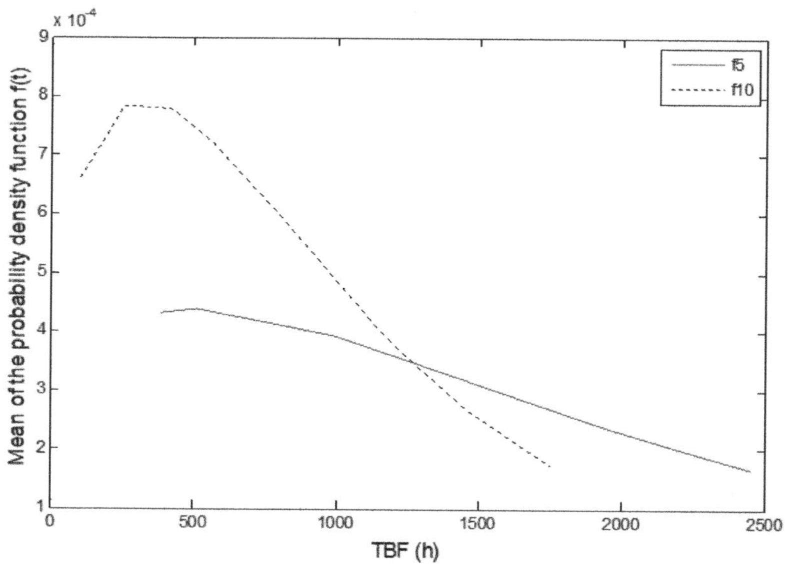

Figure 6.7. *(a) Mean of the probability density function of the machines in aging life*

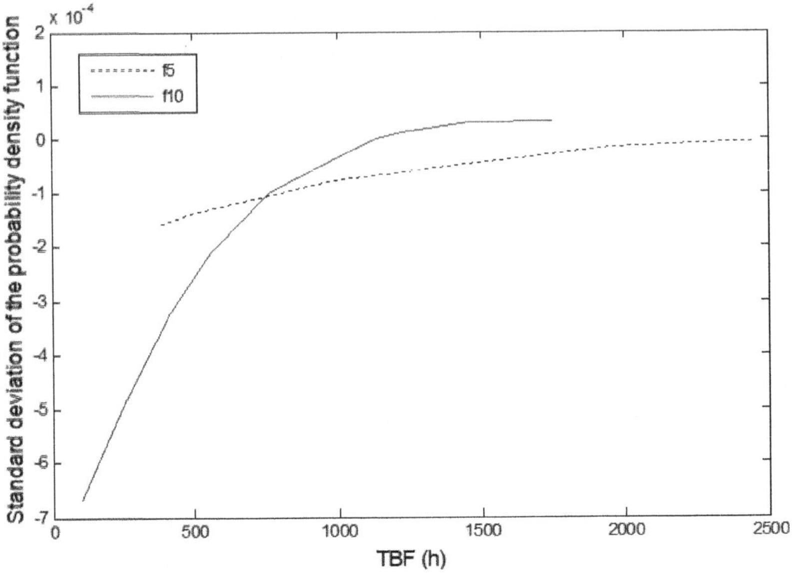

Figure 6.7. *(b) Standard deviation of the probability density function of the machines in aging life*

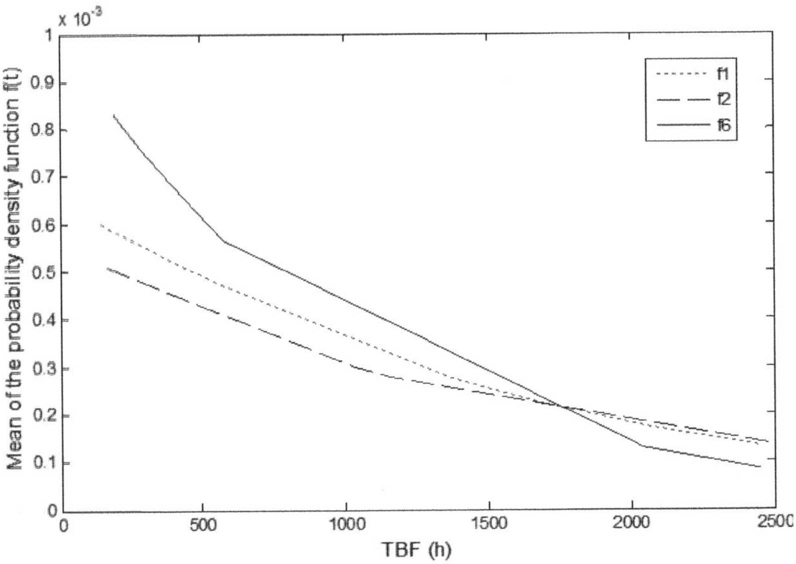

Figure 6.8. *(a) Mean of the probability density function of the machines in useful life*

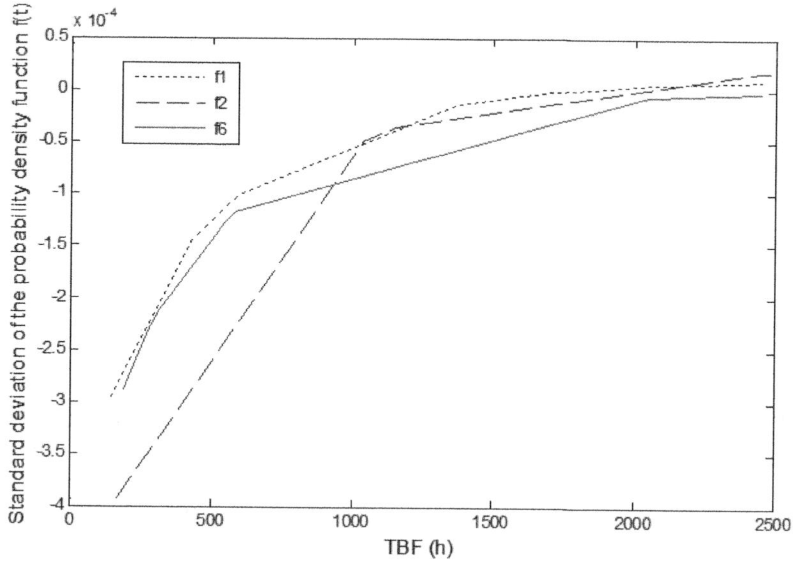

Figure 6.8. *(b) Standard deviation of the probability density function of the machines in useful life*

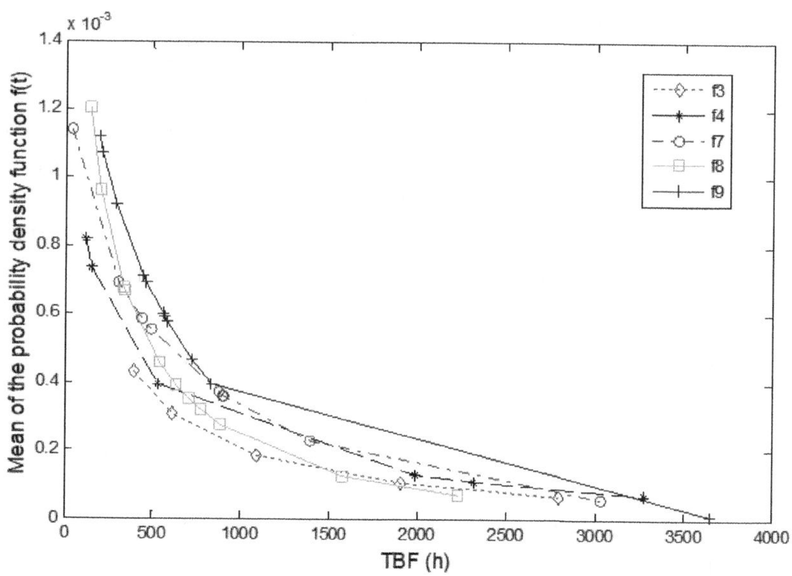

Figure 6.9. *(a) Mean of the probability density function of the machines in early life*

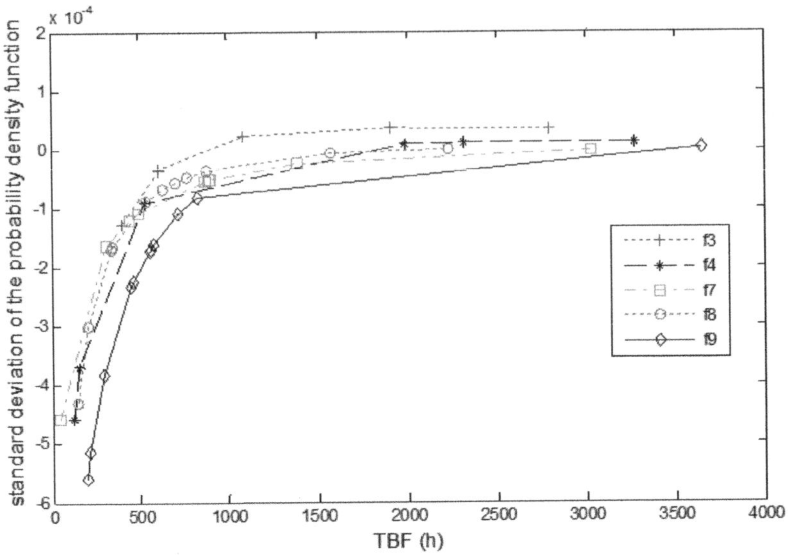

Figure 6.9. *(b) Standard deviation of the probability density function of the machines in early life*

The analysis of random reliability provides several practical recommendations for the maintenance of dialysis machines. To reduce the adverse effects of failures on the dialysis system, it is advised to enhance operational management. Furthermore, maintenance strategies should initially prioritize machines M5 and M10, which are in the aging phase. Given that dialysis machines operate in a stochastic environment with multiple sources of disturbances, it is essential to account for uncertainties throughout all calculation phases and across all variables involved.

6.3.4. *The random systematic inspection period of dialysis machines*

Continuous monitoring of failure data for the dialysis system is essential. Re-evaluating the reliability indices allows for the assessment and optimization of the maintenance strategy for the group of devices. Such practices can help reduce failures and repair times, thereby enhancing the overall efficiency of the system.

In practical terms, the objective is to ensure maximum reliability and operational safety, particularly for machines in the aging phase (M5 and M10). Experts in the field have recommended a target reliability of R = 0.95 to meet this goal. Based on the desired reliability level, the systematic inspection period (θ) for the machines can be determined using the reliability expression as a function of time, as given in equation [6.32].

$$\theta = -\eta(LogR)^{\frac{1}{\beta}} \qquad [6.32]$$

The calculations determined the systematic inspection period for the two dialysis machines in the aging phase (M5 and M10), as summarized in Table 6.4.

Devices	M5	M10
θ (h)	127	70

Table 6.4. *The systematic inspection period*

The results highlight the propagation of errors within the system and their substantial impact on reliability, especially during the aging phase. In this context, the systematic inspection period (θ) is also affected and can be represented as a Gaussian variable. The standard deviation of the systematic inspection period is expressed through the following formulation (equation [6.33]):

$$\tilde{\theta} = \theta + \sigma_\theta \varepsilon = -\tilde{\eta}(LogR)^{\frac{1}{\tilde{\beta}}} \qquad [6.33]$$

The logarithmic form of equation [6.33] leads to formulate equation [6.34]:

$$\tilde{\beta}\left[Log(-\tilde{\theta}) - Log(\tilde{\eta})\right] = Log(Log(R)) \qquad [6.34]$$

The development of the first-order Taylor series expansion of equation [6.34] is given in equation [6.35]:

$$(\beta + \sigma_\beta \varepsilon)\left[Log(-\theta) - (\sigma_\theta/\theta)\varepsilon - Log(\eta) - (\sigma_\eta/\eta)\varepsilon\right]$$

$$= Log(Log(R)) \qquad [6.35]$$

The standard deviation of the systematic inspection period is formulated through equation [6.36]:

$$\sigma_\theta = (\theta/\beta)\left[-\beta\left(\sigma_\eta/\eta\right) + \sigma_\beta\, log(-\theta) - \sigma_\beta\, log(\eta)\right] \qquad [6.36]$$

The standard deviation of the systematic inspection period (σ_θ) of M5 and M10 is given in Table 6.5.

Devices	M5	M10
σ_θ (h)	37	43

Table 6.5. *The standard deviation of the systematic inspection period*

It is noteworthy that the standard deviation confirms the substantial uncertainty impacting the systematic inspection period. For instance, in the case of machine M10, this uncertainty can exceed 60% of the mean. To ensure proper operation and availability of the machines, it is essential to account for the observed errors when determining the systematic inspection intervals.

6.4. Conclusion

This study examines the impact of stochastic Weibull parameters on the reliability of dialysis machines. The shape and scale parameters were estimated from the failure history using both analytical and graphical methods. Analytical formulations were developed and applied through numerical simulations, and the uncertainties associated with the distribution parameters were explicitly addressed. The Weibull parameters were treated as Gaussian random variables, and their means and standard deviations were calculated across multiple configurations of the dialysis devices. The failure rates and reliability of these machines in uncertain environments were analyzed for different life phases. The mean and standard deviation of the reliability indicators were derived using a first-order Taylor series expansion. Since dialysis machines operate in environments subject to multiple sources of disturbance, it is essential to account for uncertainties throughout all operational phases. Numerical results demonstrated the critical importance of considering uncertainties in the Weibull parameters, which can significantly influence machine reliability. Disturbances in the shape and scale parameters were shown to be strongly amplified in reliability indicators, posing a substantial risk to the safe and correct operation of the equipment. Moreover, the study examined the random systematic inspection period as a basis for implementing an effective monitoring and surveillance program.

Several recommendations for the maintenance department were derived from the reliability analysis in the stochastic environment. These include establishing a comprehensive safety plan, defined as a protocol that implements a series of actions to minimize errors and risks inherent in medical operations. A failure mode and effects analysis (FMEA)-based strategy should be applied, with periodic re-evaluation of failure modes and reassessment of risk priorities. The dialysis device supervisor must foster a safety culture while overseeing quality assessment and performance improvement processes. A traceability system should be employed to document maintenance actions and identify the responsible technician. Continuous training and refresher programs for personnel are crucial to enhancing machine reliability and availability. For optimal operational safety, each dialysis machine should have an individualized safety plan, taking into account its failure history, operational environment and usage practices.

It is important to note that uncertainties in the Weibull parameters propagate and amplify, potentially compromising machine reliability. Traditional reliability methods are insufficient to manage such uncertainties. Therefore, the maintenance department should explore advanced approaches to monitor and ensure machine reliability in stochastic environments. This work could be further extended by integrating artificial intelligence to provide continuous monitoring and serve as a decision-support tool for the maintenance team.

7

Fault Tree and Bayesian Network-Based Probabilistic Modeling for Predictive Maintenance of Dialysis Machines

7.1. Introduction

The maintenance of biomedical equipment, especially dialysis machines, is crucial for ensuring patient safety and minimizing service interruptions. This study aims to develop a probabilistic framework for predictive maintenance adapted to the complexity of dialysis systems. The main research question is how to improve fault detection and decision-making through integrated probabilistic tools based on clinical data. A hybrid model combining fault tree analysis (FTA) and Bayesian networks (BNs) is proposed to represent failure dependencies and quantify uncertainties. Maintenance records from dialysis machines are used to construct and validate the model. Supporting technologies such as the Internet of Medical Things (IoMT), machine learning and supervision platforms enhance data integration and real-time condition monitoring. The integrated FTA-BN approach enables early detection of faults, improved failure prediction and optimized maintenance planning. The model outperforms conventional preventive methods in identifying critical failure scenarios and supporting timely interventions. The results also emphasize the importance of high-quality data, interoperability between systems and algorithm transparency. This study demonstrates that probabilistic modeling with FTA and BNs provides a reliable approach for intelligent maintenance of dialysis machines. The framework supports a shift toward predictive, evidence-based strategies in healthcare. Ethical and regulatory challenges, including data privacy, cybersecurity and AI explainability, must be addressed to ensure safe and trusted implementation.

7.2. Literature review

The maintenance of biomedical equipment is fundamental to guaranteeing patient safety and ensuring uninterrupted healthcare delivery. Dialysis machines, due to their complexity and critical role, particularly exemplify the need for advanced maintenance strategies. Traditional maintenance approaches, often based on scheduled interventions or reactive repairs, fail to accommodate the variability of operational conditions and unexpected failure modes, leading to inefficiencies and potential risks (Lin et al. 2024). This limitation has spurred the adoption of predictive maintenance strategies, which rely on real-time data analysis to anticipate equipment failures before they occur (Sabah et al. 2022).

Predictive maintenance, supported by machine learning algorithms, enables early anomaly detection, reduced unplanned downtime and optimized resource allocation. However, its success hinges on high-quality data availability, robust computational infrastructures and rigorous validation protocols – requirements that are not easily met across all healthcare facilities (Abd Wahab et al. 2025). The IoMT plays a pivotal role in this context by enabling continuous monitoring and connectivity through smart sensors integrated into biomedical equipment. In dialysis machines, such integration supports real-time performance tracking and remote anomaly detection, enhancing maintenance scheduling and operational efficiency (Gatouillat et al. 2018). Nonetheless, these benefits come with challenges such as cybersecurity threats, data privacy concerns and interoperability issues among heterogeneous platforms (Williams and Woodward 2015).

Artificial intelligence (AI), particularly supervised and unsupervised learning models, contributes significantly to maintenance processes by classifying operational states and detecting abnormal patterns without predefined labels (Zamzam et al. 2023). These methods increase fault classification accuracy and guide targeted maintenance actions. However, the interpretability of AI models remains a concern, as their black-box nature often necessitates expert oversight in contextualizing the outputs and ensuring clinical relevance (Xu et al. 2023).

BNs offer a promising alternative for modeling uncertainty in complex healthcare systems. As probabilistic graphical models, they facilitate failure probability estimation based on historical data and observed symptoms. In dialysis machine maintenance, BNs enable dynamic diagnostic reasoning and informed decision-making (McLachlan et al. 2020). Despite their potential, these models require significant expert knowledge and computational power, particularly for updating the model as new data are introduced (Chatzimichail and Hatjimihail 2023). Their combination with other diagnostic tools can further strengthen their applicability in clinical environments (David and Jahnke 2004).

FTA, widely used in reliability engineering, complements this approach by decomposing system failures into basic events. This method aids in identifying root causes and understanding the contribution of individual components to overall system risk (Ruijters and Stoelinga 2015). In the context of dialysis equipment, FTA supports preventive maintenance by visualizing potential failure paths. However, its reliance on static and binary logic limits its effectiveness in representing uncertainty or adapting to dynamic operational conditions (Hyman and Johnson 2008).

The integration of BNs with FTA creates a hybrid modeling approach that merges the strengths of both techniques. This combination facilitates a more nuanced representation of uncertainty, enhances rare event diagnostics and supports comprehensive risk assessments at the system level (Hunte et al. 2024). However, these hybrid models are inherently complex and demand interdisciplinary collaboration to be effectively deployed in clinical settings (Leming et al. 2023). Digital twins – virtual replicas of physical assets – are also increasingly used in medical maintenance. When applied to dialysis machines, digital twins simulate equipment behavior under various scenarios, enabling predictive forecasting and proactive intervention planning (Rudsari et al. 2025). Although their advantages are substantial, digital twins face challenges such as real-time synchronization, high-fidelity modeling and seamless integration with hospital IT systems (Meijer et al. 2023).

Modern smart hospitals increasingly adopt supervision platforms based on digital dashboards that compile data from multiple sources and provide actionable insights. These platforms improve proactive maintenance and intervention traceability (Pestana et al. 2020). However, their effectiveness depends on user training, proper calibration and continuous algorithm updates to accommodate evolving device features (Sundaramurthy and Vaithiyalingam 2025; Taştan 2025).

Ethical and regulatory concerns also accompany the deployment of intelligent maintenance systems. The opacity of AI algorithms can create accountability issues, particularly when clinical decisions rely on algorithmic recommendations (Yu et al. 2024). Therefore, transparency, fairness and explainability are crucial to responsibly implementing these systems in healthcare (Bernal and Mazo 2022).

Effective implementation of intelligent maintenance strategies depends on close collaboration between biomedical engineers, clinicians and data scientists. Such interdisciplinary approaches ensure that predictive models are tailored to practical needs and meet safety standards (Tambe et al. 2025). In parallel, risk-based prioritization strategies are gaining traction for efficient resource allocation, focusing on high-failure-probability and high-criticality components (Taghipour et al. 2011). Nevertheless, these strategies require reliable risk models and frequent recalibration to remain effective in changing hospital environments (Lee et al. 2015).

Advanced monitoring systems embedded in dialysis machines enable early detection of degradation and warning signs, thus reducing treatment disruptions and extending equipment life (Jaros et al. 2023). These systems must be regularly validated and controlled, as sensor noise, calibration errors and signal drift can affect performance (Davis et al. 2020).

Fuzzy logic and neural networks are increasingly being applied in maintenance decision-making, offering flexible support tools that can handle uncertainty and imprecision in healthcare settings (David and Jahnke 2004). Despite their advantages, these methods can be computationally intensive and require specialized expertise (Wu et al. 2007).

In situations where digital infrastructure is limited, knowledge-based expert systems offer viable alternatives by relying on expert-defined rules to guide maintenance decisions. They are particularly useful in contexts lacking sufficient historical data (McCauley and Ala 1992). However, their main limitations lie in their poor scalability and their inability to learn or adapt without manual updates (Sabzaliyev 2024).

For long-term sustainability, intelligent maintenance systems must be assessed on cost-effectiveness, environmental impact and scalability. Studies have shown that predictive maintenance, through lifecycle assessments, contributes to lower operational costs and reduced resource consumption (Li et al. 2019). Still, the high initial investment and slow return on investment can be deterrents, especially in resource-limited settings (Taghipour et al. 2011).

Looking ahead, a probabilistic study based on historical failure data is planned to evaluate failure modes in dialysis machines using a combination of FTA and BNs. This integrated approach will support critical event prediction and the design of optimized maintenance plans (Li et al. 2019). It offers a robust framework for predictive maintenance tailored to hospital needs, enhancing equipment reliability and patient safety (Weber et al. 2022). Ultimately, this data-driven methodology represents a significant advancement in the digital transformation of biomedical engineering and the adoption of evidence-based maintenance practices (Mijailovic 2003).

7.3. Reliability and failure analysis of hemodialysis machines using clinical data

This section discusses the operation and significance of hemodialysis machines, which are essential for treating patients with end-stage renal disease by compensating for the kidneys' inability to remove waste and excess fluids. The process involves circulating the patient's blood through a dialyzer, where toxins are removed by diffusion and ultrafiltration across a semi-permeable membrane while preserving essential solutes. The hemodialysis machine regulates blood and dialysate flow, temperature and pressure precisely to ensure safe and effective therapy (Ronco and La Greca 2002).

The reliability of hemodialysis machines is critical for patient safety and healthcare continuity. Failures in subsystems such as sensors, blood pumps or temperature control can compromise treatment quality. Studies have demonstrated that inadequate maintenance directly correlates with reduced dialysis efficiency, emphasizing the need for regular servicing to prevent adverse outcomes. Furthermore, while the alarm system plays a crucial role in alerting staff to anomalies, it can lead to false alarms and technical disruptions if not properly calibrated (Mutia et al. 2018). Data collected from four Fresenius hemodialysis machines (M4, M5, M8 and M10) in a hospital setting between 2013 and 2022 highlight the importance of proactive maintenance strategies. These devices operated daily in a hospital setting for approximately 3.5 h each. The recorded metrics included failure counts and total repair times (Tables 7.1–7.4), providing insights into their operational reliability. Predictive maintenance techniques, which aim to detect early signs of malfunction and minimize unscheduled downtime, have been shown to enhance machine availability and reduce the risk of treatment interruptions.

Inventory No.: Machine 4		**Serial No.: NS1V5ATG46**
Failure description	$N_{Failure}$	**TTR (h)**
Flow parameter calibration problem	3	64.5
Venous pressure nozzle problem	1	5.25
Hydrophobic filter problem	1	5.25
Conductivity problem	1	7
Acid pump problem	2	10.5
Bicarbonate pump problem	1	7
Pipe problem	1	56
Filter problem	3	21

Table 7.1. *Classification of the failure history of M4 according to the nature of failure*

Inventory No.: Machine 5		**Serial No.: 1V5ATG31**
Failure description	$N_{Failure}$	**TTR (h)**
Conductivity problem	1	7
Flow parameter calibration problem	1	3.5
Venous pressure nozzle problem	1	3.5
Hydrophobic filter problem	1	3.5
Water leak problem	1	7
Filter problem	1	7
Acid pump problem	1	325

Table 7.2. *Classification of the failure history of M5 according to the nature of failure*

Inventory No.: Machine 8		**Serial No.: 1V5ATG39**
Failure description	**$N_{Failure}$**	**TTR (h)**
Bicarbonate pump problem	1	59.5
Acid pump problem	1	84
Venous pressure nozzle problem	2	5.25
Hydrophobic filter problem	2	5.25
Dialysate pump problem	4	35
Flow parameter calibration problem	2	17.5
Conductivity problem	1	7

Table 7.3. *Classification of the failure history of M8 according to the nature of failure*

Inventory No.: Machine 10		**Serial No.: 1V5ATF70**
Failure description	**$N_{Failure}$**	**TTR (h)**
Bicarbonate pump problem	1	21
Flow parameter calibration problem	1	17.5
Venous pressure nozzle problem	2	7
Hydrophobic filter problem	2	7
Conductivity problem	1	10.5
Water leak problem	1	3.5
Filter problem	2	24.5

Table 7.4. *Classification of the failure history of M10 according to the nature of failure*

In the following paragraph, the operational data will be used to construct fault tree diagrams, which systematically identify potential failure modes and their contributing causes. This approach, in line with current methodologies in biomedical engineering, aims to improve the reliability modeling of hemodialysis machines and enhance maintenance planning for critical medical equipment (Lin et al. 2022).

7.4. FTA of dialysis machines

7.4.1. *Fault tree construction*

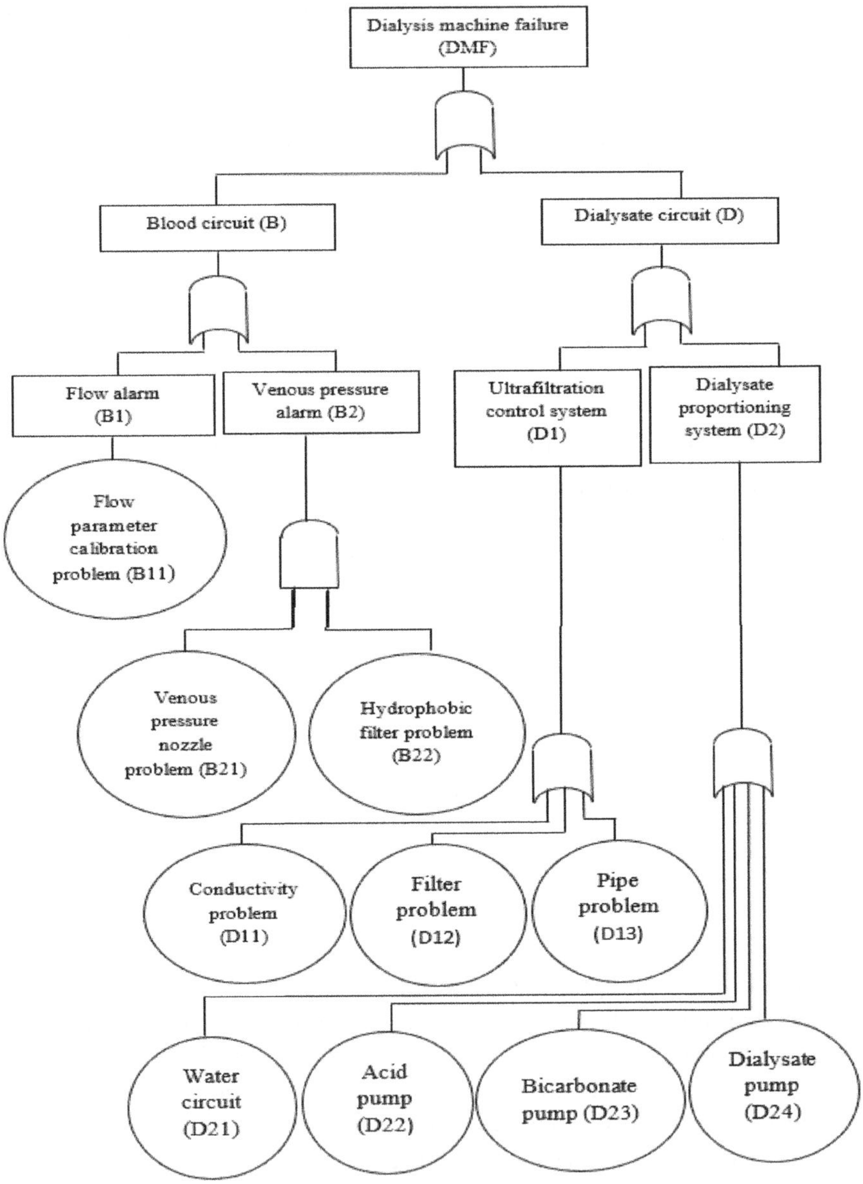

Figure 7.1. *Qualitative fault tree of the dialysis machine*

To construct the fault tree shown in Figure 7.1, the first step is to identify the top undesired event, which in this case is the overall failure of the dialysis machine. This main event is then broken down into sub-events corresponding to the major functional subsystems, such as the blood circuit, the ultrafiltration control system and the dialysate circuit. Each subsystem is analyzed in detail to determine possible root causes, such as a malfunctioning acid pump, a clogged filter or a failure in the venous pressure sensor. The tree structure relies on two types of logic gates: the OR gate, used when the occurrence of any one of the input events is sufficient to trigger the output event (for instance, a fault in the dialysate preparation system *or* in the control system can cause the machine to stop), and the AND gate, which requires the simultaneous occurrence of multiple failures to produce the undesired effect (such as a concurrent failure of the bicarbonate pump and a pipe fault). This hierarchical representation facilitates the identification of critical system components and helps prioritize risks. It also serves as a valuable tool for preventive analysis, maintenance planning and regulatory documentation. Moreover, incorporating quantitative methods such as BNs or probabilistic analyses enhances the predictive capability of the fault tree (Hunte et al. 2024), allowing for better failure management (Khan et al. 2018), a reduction in unexpected downtimes (David and Jahnke 2004) and an overall improvement in patient safety (Lodi et al. 2010), especially in sensitive medical devices like hemodialysis machines (Wu et al. 2025).

7.4.2. *Probability calculation of basic failure causes*

Table 7.5 presents the calculation of probabilities associated with basic failure causes for four dialysis machines (M4, M5, M8 and M10), based on a hierarchical fault coding system (e.g. B11 and D21). For each identified cause, the number of occurrences N_i is recorded, and a probability $F(t_i)$ is estimated using a ranking-based formula $\left(F(t_i) = \frac{i-0.3}{n+0.4}\right)$. This method relies on an empirical approach to the distribution function, commonly used in failure analysis within medical environments (Tsarouhas 2019). The shaded areas in the table indicate the absence of certain causes for specific machines. This type of analysis helps prioritize maintenance actions by highlighting the most recurrent failures (Tsai et al. 2025). Furthermore, the results can be integrated into intelligent diagnostic systems to enhance the reliability and availability of dialysis equipment, which is essential for ensuring continuity of care (Kotanko et al. 2023).

Failure code			M4		M5		M8		M10	
			N_i	F(ti)	Ni	F(ti)	N_i	F(ti)	N_i	F(ti)
B	**B1**	**B11**	3	0.201	1	0.094	2	0.126	1	0.067
	B2	**B21**	1	0.052	1	0.094	2	0.126	2	0.163
		B22	1	0.052	1	0.094	2	0.126	2	0.163
D	**D1**	**D11**	1	0.052	1	0.094	1	0.126	1	0.067
		D12	3	0.201	1	0.094			2	0.163
		D13	1	0.052						
	D2	**D21**			1	0.094			1	0.067
		D22	2	0.126	1	0.094	1	0.126		
		D23	1	0.052			1	0.126	1	0.067
		D24					4	0.276		
			$N_{total} = 13$		$N_{total} = 7$		$N_{total} = 13$		$N_{total} = 10$	

Table 7.5. *Probability calculation of basic failure causes*

7.4.3. Probability calculation of intermediate failure events

Table 7.6 presents the calculation of the probabilities of intermediate failure events for four dialysis machines (M4, M5, M8 and M10), as part of the construction of a fault tree. Each intermediate event (such as B, D1, D2 and D) is derived by combining the probabilities of the basic failure causes previously identified (Table 7.5). The formulas shown are based on Boolean algebra rules: events connected by OR gates (e.g. $P(B) = P(B1)+P(B2)$) are summed, while those connected by AND gates (e.g. $P(B2) = P(B21) \times P(B22)$) are multiplied (Vesely et al. 1981). Finally, the overall probability of a major functional failure (dialysis machine failure (DMF)) is obtained by summing the probabilities of the two main branches, B and D. This hierarchical method provides a structured assessment of reliability, which is essential in biomedical maintenance (Hyman and Johnson 2008).

Failure code	Probability formula	M4	M5	M8	M10
B1	$P(B1) = P(B11)$	0.201	0.094	0.126	0.067
B2	$P(B2) = P(B21) \times P(B22)$	0.002	0.008	0.015	0.026
B	$P(B) = P(B1) + P(B2)$	0.203	0.102	0.141	0.093
D1	$P(D1) = P(D11) + P(D12) + P(D13)$	0.305	0.188	0.052	0.230
D2	$P(D2) = P(D21) + P(D22) + P(D23) + P(D24)$	0.178	0.188	0.38	0.134
D	$P(D) = P(D1) + P(D2)$	0.483	0.376	0.432	0.364
DMF	$P(\text{DMF}) = P(B) + P(D)$	**0.686**	**0.478**	**0.573**	**0.457**

Table 7.6. *Calculation of probabilities of intermediate events of failure of machines*

7.5. BN analysis of dialysis machines

7.5.1. *BN construction*

The transformation of a fault tree into a BN enhances reliability analysis by incorporating uncertainties and dependencies between events (Figure 7.2). This process begins by identifying the basic and intermediate events from the fault tree, each of which becomes a node in the directed graph of the BN. Logical relationships such as AND and OR gates are then translated into causal links between nodes. For instance, an AND gate becomes a conjunctive dependency where the occurrence of an event depends on all its predecessors, while an OR gate implies that the occurrence of at least one parent event is sufficient to trigger the target event (Weber et al. 2022). Each node is associated with a conditional probability table (CPT), constructed based on the logical rules of the original tree.

The main advantage of a BN lies in its ability to perform belief propagation, allowing dynamic updating of probabilities based on available observations. This analytical flexibility is particularly relevant for critical medical devices, where operating conditions can vary and data may be incomplete (Langseth and Portinale 2007).

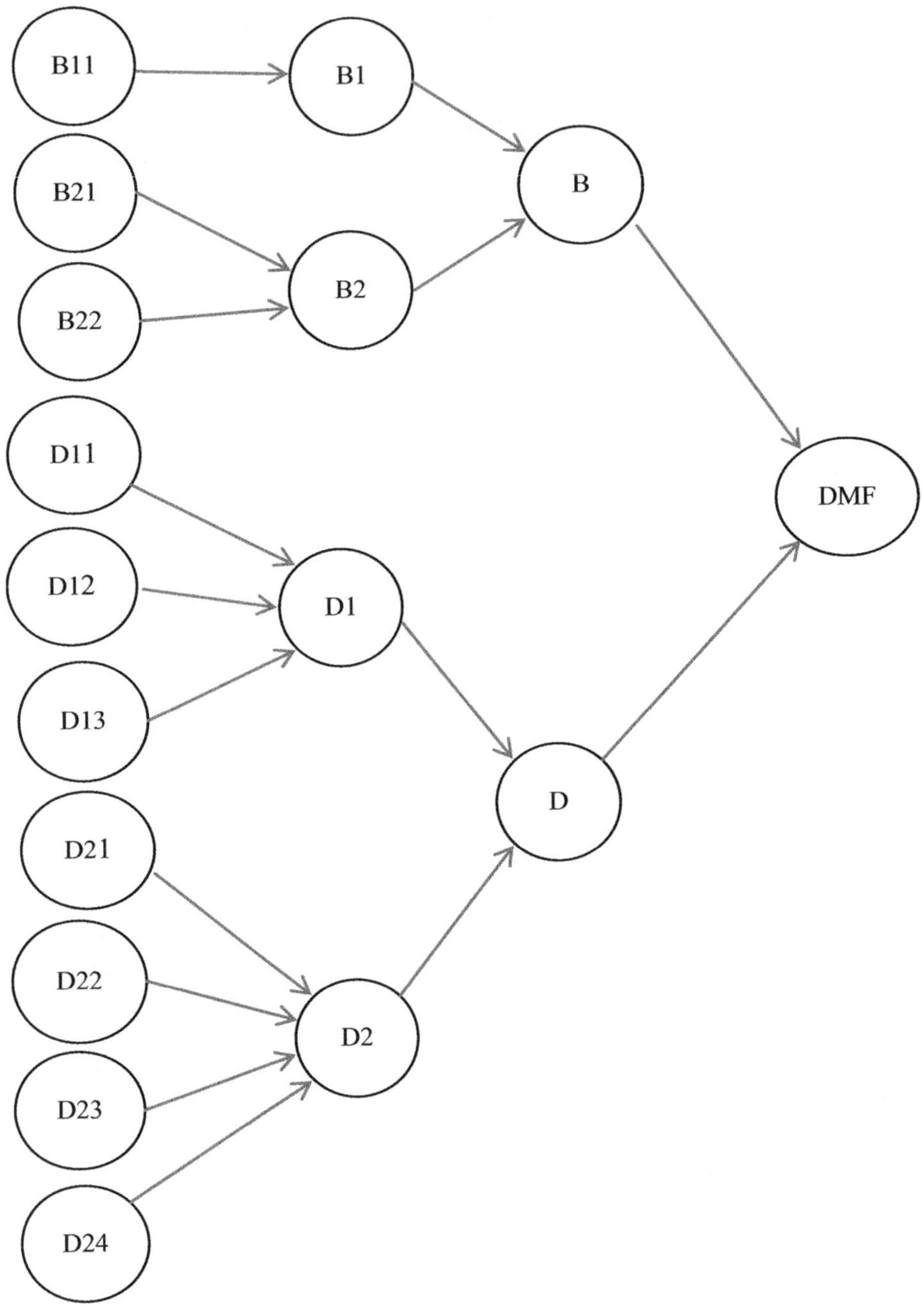

Figure 7.2. *Structure of the Bayesian network obtained from the fault tree*

7.5.2. Probability calculation of intermediate failure events

On the basis of Table 7.5, which provides the probabilities of basic failure causes (noted as $F(t_i)$) for different dialysis machines (M4, M5, M8 and M10), the probabilities of intermediate failure events can be calculated using a BN. The first step involves identifying the basic events B_{ij}, D_{ij}, etc., and extracting their individual probabilities from the observed frequencies in Table 7.5. These probabilities are then combined according to the logical relationships shown in the truth tables (Table 7.7), where intermediate events (such as B, D_1, D_2, D and DMF) are expressed using logical functions of the basic events. For each combination of these events, a joint probability is calculated using the conditional probability rules specific to BNs.

For example, the probability of event D_1 is determined by the combined contributions of events D_{11}, D_{12} and D_{13}, following an OR–AND logic formula detailed in the tables. These relationships are then used to compute the probability of D, integrating the probabilities of D_1 and D_2, and ultimately leading to the probability of the major failure event DMF, which is the final objective of the model. The numerical results are presented in Table 7.8.

The purpose of this calculation is to quantify the overall failure risk of a machine based on elementary failures, even in the presence of uncertainty. The Bayesian approach allows the updating of probabilities as new data becomes available, enabling proactive maintenance decision-making. The goal is therefore to detect critical failures, prioritize maintenance actions and improve the reliability and safety of biomedical equipment.

B21	0		1	
B22	0	1	0	1
B2	0	0	0	1

$$P(B2) = P(B21) \times P(B22) \quad [7.1]$$

D11	0				1			
D12	0		1		0		1	
D13	0	1	0	1	0	1	0	1
D1	0	1	1	1	1	1	1	1

$$P(D1) = [(1 - P(D11)) \times (1 - P(D12)) \times P(D13)] + [(1 - P(D11)) \times P(D12)] + P(D11) \quad [7.2]$$

D21	0								1							
D22	0				1				0				1			
D23	0		1		0		1		0		1		0		1	
D24	0	1	0	1	0	1	0	1	0	1	0	1	0	1	0	1
D2	0	1	1	1	1	1	1	1	1	1	1	1	1	1	1	1

$$P(D2) = \left[\left(1 - P(D21)\right) \times \left(1 - P(D22)\right) \times \left(1 - P(D23)\right) \times P(D24)\right] + \left[\left(1 - P(D21)\right) \times \left(1 - P(D22)\right) \times P(D23)\right] + \left[\left(1 - P(D21)\right) \times P(D22)\right] + P(D21) \quad [7.3]$$

B1	0		1	
B2	0	1	0	1
B	0	1	1	1

$$P(B) = \left[\left(1 - P(B1)\right) \times P(B2)\right] + P(B1) \quad [7.4]$$

D1	0		1	
D2	0	1	0	1
D	0	1	1	1

$$P(D) = \left[\left(1 - P(D1)\right) \times P(D2)\right] + P(D1) \quad [7.5]$$

B	0		1	
D	0	1	0	1
DMF	0	1	1	1

$$P(DMF) = \left[\left(1 - P(B)\right) \times P(D)\right] + P(B) \quad [7.6]$$

Table 7.7. *Calculation of conditional probability of dialysis machine*

Failure code	**M4**	**M5**	**M8**	**M10**
B1	0.201	0.094	0.126	0.067
B2	0.002	0.008	0.015	0.026
B	0.202	0.101	0.139	0.091
D1	0.282	0.179	0.126	0.219
D2	0.171	0.179	0.349	0.129
D	0.404	0.326	0.431	0.320
DMF	0.524	0.394	0.510	0.382

Table 7.8. *Probabilities of intermediate events of failure of dialysis machines through Bayesian method*

7.6. Interpretation of results

The use of BNs to calculate inverse conditional probabilities is a powerful and comprehensive method for diagnosing failures and improving the reliability of complex systems such as dialysis machines. Traditional FTA is limited in that it only allows the calculation of the overall failure probability of a system based on the logical combination of basic events (e.g. sensor failures, valve blockages or software errors), assuming their independence. While FTA can identify potential failure paths, it does not provide detailed posterior probabilities once a failure is observed.

The combined use of the fault tree and the BN in the previous study makes it possible to precisely identify the most critical combinations of basic events leading to a major failure of the dialysis system (DMF). For example, according to Table 7.8, the overall failure probability for machine M4 is 0.524, the highest among the four machines analyzed. By examining the contribution of subsystems to this probability, it is observed that events D1 and D2, associated with internal components of module D, have probabilities of 0.282 and 0.171, respectively. These results indicate that module D is a vulnerable area that requires special attention in terms of maintenance. Tracing back through the fault tree, it becomes evident that event D1 is strongly influenced by the sub-event D12, which has a probability of 0.201 (Table 7.5), making it a high-priority target for corrective actions.

Moreover, the analysis identifies the most dangerous basic failure, meaning the one that, when it occurs, has the highest conditional probability of leading to a major system failure. According to the results obtained from the BN, this failure is associated with event D12. In fact, when the global failure DMF is observed, the probability that D12 is the cause reaches 0.689, making it the event most strongly correlated with total system failure. This information is essential for guiding maintenance efforts toward the most decisive causes.

To understand this process, let us start with a simplified analytical computation using Bayes' theorem:

$$P(D12 = 1/\mathrm{DMF} = 1)$$

$$= P(\mathrm{DMF} = 1/D12 = 1) \times P(D12 = 1)/P(\mathrm{DMF} = 1) \qquad [7.7]$$

Assuming the following values:

$$P(D12 = 1) = 0.201$$

$\mathrm{P(DMF = 1/D12 = 1) = 0.9}$: the probability that the system fails if D12 is faulty.

$$P(\mathrm{DMF} = 1) = 0.524$$

We compute:

$$P(D12 = 1/\mathrm{DMF} = 1) = 0.9 \times 0.201/0.524 = 0.345 \quad [7.8]$$

This intermediate result (0.383) gives an estimate of the probability that D12 caused the failure, based only on its direct influence on DMF. However, this simplified method does not take into account the influence of other components or the full network dependencies.

To get a more accurate and global estimation, we apply full Bayesian inference across the network. Using the joint probability distribution of the network, the posterior is computed by marginalizing over all possible configurations of the other variables:

$$P(D12 = 1/\mathrm{DMF} = 1) =$$

$$P(D12 = 1, \mathrm{DMF} = 1)/P(\mathrm{DMF} = 1) = 0.361/0.524 = 0.689 \quad [7.9]$$

This value (0.689) means that once the system failure DMF is observed, there is a 68.9% chance that the component D12 is responsible. It becomes the most likely root cause. This is significantly higher than the initial probability of D12 failing (0.201), showing how evidence (DMF = 1) drastically shifts the belief about the source of the problem.

This value is given through the global inference mechanism of the BN, which considers all interdependencies among components and uses exact inference algorithms (e.g. belief propagation) to update beliefs based on observed evidence.

However, discrepancies may arise between the results obtained from the classical fault tree and those from the BN. These differences mainly stem from the way uncertainties and dependencies between events are handled.

The fault tree generally relies on simplifying assumptions such as the independence of basic events and the use of static logic gates (AND/OR), which can lead to overestimation or underestimation of risks. In contrast, the BN incorporates the notion of conditional probability, allowing for a more detailed modeling of interactions between events and dynamically updating estimates as new data becomes available. For example, if information is known about a partial event already occurring (such as a failure in D12), the BN automatically adjusts the probability of DMF, which is not possible with the fault tree approach.

Thus, the differences between the two methods highlight the value of the BN for a more realistic and dynamic analysis, promoting better prioritization of interventions. With this combined approach, it becomes possible not only to identify critical points but also to assess the impact of targeted maintenance strategies. This leads to improved system reliability while optimizing available resources in a data-driven preventive maintenance logic.

In conclusion, BNs enhance the classical approach by enabling reasoning under uncertainty, prioritizing components based on observed failures and continuously adapting as new evidence is introduced. The ability to compute P(Component = 1 | System failure = 1) allows engineers not only to react faster but also to predict, prevent, and optimize maintenance actions. This represents a major step forward in the evolution from reactive to predictive and evidence-based maintenance strategies.

7.7. Conclusion

This chapter presents a methodology that combines FTA with BNs to provide a more accurate assessment of the reliability of a complex system, specifically a dialysis machine.

The initial phase consisted of modeling the system using a fault tree, which allowed for the logical identification of various combinations of basic events, which could lead to a major failure. This structure was then converted into a BN to incorporate uncertainties, account for dependencies between events and dynamically update probabilities as new observations become available.

One of the key advantages of this approach lies in its ability to calculate conditional probabilities. It enables the estimation of the likelihood that a specific component is faulty when a system-wide failure is observed, or conversely, to evaluate how a local failure could influence the overall behavior of the system. This provides a more realistic and flexible reliability analysis, overcoming the limitations of traditional models that rely on static logic and assumptions of independence between events.

The results made it possible to prioritize events based on their actual impact on system performance, which is crucial for optimizing maintenance planning. This allows maintenance teams to focus efforts and resources on the most critical components, anticipate failures more effectively and ultimately improve equipment availability.

The prospects of this work are particularly promising. The proposed methodology can be extended to other critical biomedical devices to support the development of predictive maintenance strategies across healthcare institutions. Furthermore, the integration of real-time data through connected sensors would allow continuous updates to the BNs, enabling early detection of anomalies and timely decision-making. Lastly, combining this approach with AI, especially machine learning techniques, could make it possible to automatically build and refine network structures using historical data, further improving predictive capabilities. This synergy between probabilistic modeling, connectivity and learning opens the way to smarter, more responsive, and more cost-efficient maintenance systems.

Conclusion

The increasing reliance on sophisticated biomedical equipment in modern healthcare institutions places an imperative on robust maintenance strategies that ensure equipment reliability, availability and safety. This work has addressed these challenges through a structured and multi-dimensional approach, focusing particularly on dialysis machines – devices whose uninterrupted performance is critical for patient survival.

Chapter 1 laid the groundwork by detailing the principles and practices involved in the management of medical equipment within healthcare facilities. It emphasized the strategic importance of integrating technical, organizational and regulatory dimensions to guarantee effective asset lifecycle management and resource optimization.

In Chapter 2, the technical and functional characteristics of dialysis machines were examined in depth. This analysis highlighted the complexity and vulnerability of these systems, particularly regarding their dependence on multiple subsystems such as fluid control, filtration and electronic monitoring – each of which is subject to wear, malfunction and environmental influence.

Chapter 3 introduced the Weibull distribution as a flexible and widely used statistical tool for modeling time-to-failure data. This distribution was presented not only as a means of understanding failure behavior across different life stages of a medical device but also as a foundation for predictive analysis.

Building on this, Chapter 4 applied the Weibull distribution to assess the reliability, maintainability and availability (RMA) of generic biomedical equipment. This quantitative analysis demonstrated how historical failure data can be transformed into actionable indicators that inform optimal maintenance intervals, reduce unplanned downtime and improve cost efficiency.

Chapter 5 focused the RMA analysis specifically on dialysis machines, leveraging real-world failure data. Through parameter estimation and Weibull modeling, it became possible to identify the most failure-prone components, forecast failure probabilities and recommend component-specific maintenance actions. This chapter thus translated statistical findings into practical maintenance improvements tailored to a life-critical device.

Chapter 6 extended the previous Weibull analysis by examining the impact of stochastic Weibull parameters on dialysis machine reliability. Instead of fixed values, the parameters were treated as random variables to capture uncertainty in failure behavior. The chapter estimated random failure rate, reliability, and probability density functions, and determined a random systematic inspection period to support more robust maintenance planning.

Finally, Chapter 7 introduced a probabilistic modeling framework for predictive maintenance using fault tree analysis (FTA) and Bayesian networks (BNs). This dual approach allowed for a deeper understanding of causal relationships between component failures and overall system risk. FTA provided a deductive, hierarchical model of failure scenarios, while BNs introduced probabilistic reasoning and dynamic data integration, enabling the continuous updating of failure predictions based on new information and real-time monitoring.

In summary, this comprehensive study proposes a transition from reactive or merely preventive maintenance strategies to a reliability-centered, data-driven and probabilistically informed maintenance model. By integrating historical failure data, statistical analysis via Weibull distribution, system-level modeling through FTA and adaptive reasoning via BNs, this approach supports the development of predictive maintenance frameworks that are both technically robust and operationally feasible.

Such a framework is particularly valuable in resource-constrained healthcare environments, where optimizing equipment uptime directly translates into better clinical outcomes and more efficient resource use. In the specific context of dialysis machines, this methodology ensures not only continuity of care but also enhanced patient safety, lower maintenance costs and extended equipment lifespan.

Ultimately, this work demonstrates that by leveraging engineering tools, statistical models and AI-driven decision systems, healthcare institutions can evolve toward a new paradigm of biomedical equipment maintenance – one that transforms data into foresight, and foresight into safety and sustainability.

APPENDICES

Appendix 1

Table for Gamma Function Γ

n	$\Gamma(n)$	n	$\Gamma(n)$	n	$\Gamma(n)$	n	$\Gamma(n)$
1.00	1.00000	1.25	0.90640	1.50	0.88623	1.75	0.91906
1.01	0.99433	1.26	0.90440	1.51	0.88659	1.76	0.92137
1.02	0.98884	1.27	0.90250	1.52	0.88704	1.77	0.92376
1.03	0.98355	1.28	0.90072	1.53	0.88757	1.78	0.92623
1.04	0.97844	1.29	0.89904	1.54	0.88818	1.79	0.92877
1.05	0.97350	1.30	0.89747	1.55	0.88887	1.80	0.93138
1.06	0.96874	1.31	0.89600	1.56	0.88964	1.81	0.93408
1.07	0.96415	1.32	0.89464	1.57	0.89049	1.82	0.93685
1.08	0.95973	1.33	0.89338	1.58	0.89142	1.83	0.93969
1.09	0.95546	1.34	0.89222	1.59	0.89243	1.84	0.94261
1.10	0.95135	1.35	0.89115	1.60	0.89352	1.85	0.94561
1.11	0.94739	1.36	0.89018	1.61	0.89468	1.86	0.94869
1.12	0.94359	1.37	0.88931	1.62	0.89592	1.87	0.95184
1.13	0.93993	1.38	0.88854	1.63	0.89724	1.88	0.95507
1.14	0.93642	1.39	0.88785	1.64	0.89864	1.89	0.95838
1.15	0.93304	1.40	0.88726	1.65	0.90012	1.90	0.96177
1.16	0.92980	1.41	0.88676	1.66	0.90167	1.91	0.96523
1.17	0.92670	1.42	0.88636	1.67	0.90330	1.92	0.96878
1.18	0.92373	1.43	0.88604	1.68	0.90500	1.93	0.97240
1.19	0.92088	1.44	0.88580	1.69	0.90678	1.94	0.97610
1.20	0.91817	1.45	0.88565	1.70	0.90864	1.95	0.97988
1.21	0.91558	1.46	0.88560	1.71	0.91057	1.96	0.98374
1.22	0.91311	1.47	0.88563	1.72	0.91258	1.97	0.98768
1.23	0.91075	1.48	0.88575	1.73	0.91466	1.98	0.99171
1.24	0.90852	1.49	0.88595	1.74	0.91683	1.99	0.99581
						2.00	1.00000

Appendix 2

One-Sample Kolmogorov–Smirnov Table

The following table gives the critical values $D_{n,\alpha}$ as described in the Kolmogorov–Smirnov test.

n \ α	**0.01**	**0.05**	**0.1**	**0.15**	**0.2**
1	0.995	0.975	0.950	0.925	0.900
2	0.929	0.842	0.776	0.726	0.684
3	0.828	0.708	0.642	0.597	0.565
4	0.733	0.624	0.564	0.525	0.494
5	0.669	0.565	0.510	0.474	0.446
6	0.618	0.521	0.470	0.436	0.410
7	0.577	0.486	0.438	0.405	0.381
8	0.543	0.457	0.411	0.381	0.358
9	0.514	0.432	0.388	0.360	0.339
10	0.490	0.410	0.368	0.342	0.322
11	0.468	0.391	0.352	0.326	0.307
12	0.450	0.375	0.338	0.313	0.295
13	0.433	0.361	0.325	0.302	0.284
14	0.418	0.349	0.314	0.292	0.274

15	0.404	0.338	0.304	0.283	0.266
16	0.392	0.328	0.295	0.274	0.258
17	0.381	0.318	0.286	0.266	0.250
18	0.371	0.309	0.278	0.259	0.244
19	0.363	0.301	0.272	0.252	0.237
20	0.356	0.294	0.264	0.246	0.231
25	0.320	0.270	0.240	0.220	0.210
30	0.290	0.240	0.220	0.200	0.190
35	0.270	0.230	0.210	0.190	0.180
40	0.250	0.210	0.190	0.180	0.170
45	0.240	0.200	0.180	0.170	0.160
50	0.230	0.190	0.170	0.160	0.150
OVER 50	$\frac{1.63}{\sqrt{n}}$	$\frac{1.36}{\sqrt{n}}$	$\frac{1.22}{\sqrt{n}}$	$\frac{1.14}{\sqrt{n}}$	$\frac{1.07}{\sqrt{n}}$

Appendix 3

The Gaussian Distribution

A3.1. Introduction

Probably the most-important distribution in all of statistics is the Gaussian distribution, also called the normal distribution. The Gaussian distribution arises in many contexts and is widely used for modeling continuous random variables.

The general form of its probability density function is given as:

$$f(x) = \frac{1}{\sqrt{2\pi\sigma^2}} exp(-\frac{(x-\mu)^2}{2\sigma^2}) \quad [A3.1]$$

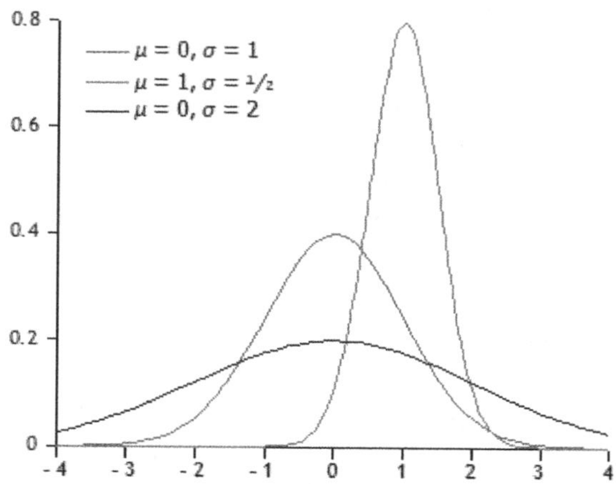

Figure A3.1. *Probability density function. For a color version of this figure, see www.iste.co.uk/bouchoucha/optimizing.zip*

The parameter μ is the mean of the distribution, while the parameter σ2 is the variance. The standard deviation of the distribution is σ. A random variable with a Gaussian distribution is said to be normally distributed, and is called a normal deviate.

Figure A3.1 plots the probability density function for several sets of parameters (μ, σ2). The distribution is symmetric around the mean and most of the density (99.7%) is contained within 3σ of the mean.

A3.2. Definitions

A3.2.1. *Standard normal distribution*

The simplest case of a normal distribution is known as the standard normal distribution or unit normal distribution. This is a special case when μ = 0 and σ = 1, and it is described by this probability density function (or density):

$$\varphi(z) = \frac{1}{\sqrt{2\pi}} exp(-\frac{z^2}{2}) \quad [A3.2]$$

The variable z has a mean of 0 and a variance and standard deviation of 1.

z is a reduced (σ =1) centered (μ=0) Gaussian variable.

A3.2.2. *General normal distribution*

Every normal distribution is a version of the standard normal distribution, whose domain has been stretched by a factor σ (the standard deviation) and then translated by μ (the mean value):

$$f(x|\mu, \sigma^2) = \frac{1}{\sigma}\varphi(\frac{x-\mu}{\sigma}) \quad [A3.3]$$

If Z is a standard normal deviate, then $X = \sigma Z + \mu$ will have a normal distribution with expected value μ and standard deviation σ. This is equivalent to saying that the standard normal distribution Z can be scaled/stretched by a factor of σ and shifted by μ to yield a different normal distribution, called X. Conversely, if X is a normal deviate with parameters

μ and σ^2, then this X distribution can be re-scaled and shifted via the formula $Z = \frac{x-\mu}{\sigma}$ to convert it to the standard normal distribution. This variate is also called the standardized form of X.

Appendix 4

Taylor Series

A4.1. Series representation of a function

A series writes a given complicated quantity as an infinite sum of simple terms. To approximate the quantity, we take only the first few terms of the series, dropping the later terms which give progressively smaller corrections.

In this section, we finally develop the tool that lets us do this in most cases: a way to write any reasonable function $f(x)$ as an explicit power series, a kind of infinite polynomial. This will allow us to compute outputs of the function by plugging values of x into the series.

Our functions must behave decently near the center point of the desired power series. We say $f(x)$ is analytic at $x = a$ if it is possible to write:

$$f(x) = \sum_{n=0}^{\infty} c_n (x-a)^n \quad \text{[A4.1]}$$

for some coefficients c_n, with positive radius of convergence.

In practice, any formula involving standard functions and operations defines an analytic function, provided the formula gives real number values in a small interval around $x = a$. For example, $\frac{1}{x-a}$ is not analytic at $x = a$, because it gives $\pm\infty$ at $x = a$; and $\sqrt{x-a}$ is not analytic at $x = a$ because for x slightly smaller than a, it gives the square root of a negative number.

A4.2. Taylor series theorem

Let $f(x)$ be a function, which is analytic at $x = a$. Then we can write $f(x)$ as the following power series, called the Taylor series of $f(x)$ at $x = a$:

$$f(x) = f(a) + f'(a)(x-a) + \frac{f''(a)}{2!}(x-a)^2 + \frac{f'''(a)}{3!}(x-a)^3 + \cdots \qquad [A4.2]$$

valid for x within a radius of convergence $|x-a| < R$ with R > 0, or convergent for all x.

If we write the nth derivative of $f(x)$ as $f^{(n)}(x)$, this becomes:

$$f(x) = \sum_{n=0}^{\infty} c_n (x-a)^n \qquad [A4.3]$$

where the coefficients $c_n = \frac{f^{(n)}(a)}{n!}$

A4.2.1. *The Taylor series development of the logarithmic function*

Applying the Taylor series formula to the logarithmic function near zero allows us to write:

$$Ln(1+x) = x - \frac{x^2}{2} + \frac{x^3}{3} - \frac{x^4}{4} + \cdots + (-1)^n \frac{x^{n+1}}{n+1} + \cdots \qquad [A4.4]$$

$$Ln(1+x) = \sum_{n=0}^{\infty} (-1)^n \frac{x^{n+1}}{n+1} \; for \; -1 < x < 1 \qquad [A4.5]$$

A4.2.2. *The Taylor series development of the exponential function*

Applying the Taylor series formula to the logarithmic function near zero allows us to write:

$$e^x = 1 + x + \frac{x^2}{2!} + \frac{x^3}{3!} - \frac{x^4}{4!} + \cdots \qquad [A4.6]$$

$$e^x = \sum_{n=0}^{\infty} \frac{x^n}{n!} \text{ for all } x \qquad [A4.7]$$

A4.3. Determination of the mean and standard deviation of the failure rate through the first-order Taylor series expansion

The random failure rate is expressed as a random function following a Gaussian distribution as follows:

$$\tilde{\lambda}(t) = \lambda(t) + \sigma_{\lambda}(t)\varepsilon \qquad [A4.8]$$

where $\lambda(t)$ is the mean of the failure rate and $\sigma_{\lambda}(t)$ its standard deviation.

According to the two-parameter Weibull distribution, the random failure rate is offered, as a function of the shape and scale parameter, by the following expression:

$$\tilde{\lambda}(t) = \frac{\tilde{\beta}}{\tilde{\eta}}\left(\frac{t}{\tilde{\eta}}\right)^{\tilde{\beta}-1} \qquad [A4.9]$$

In order to formulate the mean and the standard deviation of the failure rate, the logarithmic failure rate can be used as follows:

$$ln\left(\tilde{\lambda}(t)\right) = \left(ln(\tilde{\beta}) - ln(\tilde{\eta})\right) + (\tilde{\beta} - 1)\left(ln(t) - ln(\tilde{\eta})\right) \qquad [A4.10]$$

$$ln\left(\tilde{\lambda}(t)\right) = ln(\tilde{\beta}) - \tilde{\beta}\, ln(\tilde{\eta}) + (\tilde{\beta} - 1)\, ln(t) \qquad [A4.11]$$

To linearize the failure rate equation, we will use the first-order Taylor series expansion of the random quantity of $\ln(\lambda(t))$ at the vicinity of its mean (λ), and we obtain:

$$ln(\tilde{\lambda}) = ln(\lambda + \sigma_{\lambda}\varepsilon) \qquad [A4.12]$$

To apply the Taylor Series Theorem, we will consider:

$$\begin{cases} f(x) = ln(x) \\ f'(x) = \frac{1}{x} \\ x = \tilde{\lambda} \\ a = \lambda \\ x - a = \tilde{\lambda} - \lambda = \sigma_{\lambda}\varepsilon \end{cases} \qquad [A4.13]$$

We therefore obtain the first-order Taylor series expansion of $\ln(\tilde{\lambda})$:

$$ln(\tilde{\lambda}) = ln(\lambda) + \frac{\sigma_\lambda}{\lambda}\varepsilon \quad [A4.14]$$

The same development for $\ln(\tilde{\beta})$ and $\ln(\tilde{\eta})$ leads to write:

$$ln(\tilde{\beta}) = ln(\beta) + (\sigma_\beta/\beta)\,\varepsilon \quad [A4.15]$$

$$ln(\tilde{\eta}) = ln(\eta) + (\sigma_\eta/\eta)\,\varepsilon \quad [A4.16]$$

A4.4. Determination of the mean and standard deviation of the reliability R(t) through the first-order Taylor series expansion

The random reliability is expressed as a random function following a Gaussian distribution as follows:

$$\tilde{R}(t) = R(t) + \sigma_R(t)\varepsilon \quad [A4.17]$$

where R(t) is the mean of the failure rate and $\sigma_R(t)$ its standard deviation.

According to the two-parameter Weibull distribution, the random reliability is offered, as a function of the shape and scale parameter, by the following expression:

$$\tilde{R}(t) = exp\,(-t/\tilde{\eta})^{\tilde{\beta}} \quad [A4.18]$$

To apply the Taylor Series Theorem, we will consider:

$$\begin{cases} f(x) = e^x \\ x = (-t/\tilde{\eta})^{\tilde{\beta}} \end{cases} \quad [A4.19]$$

The first-order Taylor series expansion of $\tilde{R}(t)$ leads us to:

$$\tilde{R}(t) = exp(-t/\tilde{\eta})^{\tilde{\beta}} = 1 + (-t/\tilde{\eta})^{\tilde{\beta}} \quad [A4.20]$$

or even:

$$\tilde{R}(t) - 1 = (-t/\tilde{\eta})^{\tilde{\beta}} \quad [A4.21]$$

The logarithmic reliability is formulated in the following equation:

$$Ln\left(\tilde{R}(t)-1\right)=ln\left[(-t/\tilde{\eta})^{\tilde{\beta}}\right]=\tilde{\beta}(Ln(-t)-Ln(\tilde{\eta})) \quad \text{[A4.22]}$$

To linearize the random logarithmic reliability equation, we will use the first-order Taylor series expansion in the vicinity of the mean R(t), so we will consider:

$$\begin{cases} f(x)=ln(x-1) \\ f'(x)=\frac{1}{x-1} \\ x=\tilde{R}(t) \\ a=R(t) \\ x-a=\tilde{R}(t)-R(t)=\sigma_R\varepsilon \end{cases} \quad \text{[A4.23]}$$

We therefore obtain the first-order Taylor series expansion of $Ln\left(\tilde{R}(t)-1\right)$:

$$Ln\left(\tilde{R}-1\right)=Ln(R-1)+\frac{\sigma_R}{R-1}\varepsilon \quad \text{[A4.24]}$$

The first-order Taylor series expansion of $\tilde{\beta}(Ln(-t)-Ln(\tilde{\eta}))$ can be written using the first-order Taylor series expansion of the logarithmic function, as follows:

$$\tilde{\beta}\left(Ln(-t)-Ln(\tilde{\eta})\right)=(\beta+\sigma_\beta\varepsilon)\left(Ln(-t)-Ln(\eta)-\frac{\sigma_\eta}{\eta}\varepsilon\right) \quad \text{[A4.25]}$$

$$\tilde{\beta}\left(Ln(-t)-Ln(\tilde{\eta})\right)=$$

$$\beta\left(Ln(-t)-Ln(\eta)\right)+\left(\sigma_\beta\left(Ln(-t)-Ln(\eta)\right)-\beta\frac{\sigma_\eta}{\eta}\right)\varepsilon \quad \text{[A4.26]}$$

$$\tilde{\beta}\left(Ln(-t)-Ln(\tilde{\eta})\right)=\beta Ln\left(\frac{-t}{\eta}\right)+\left(\sigma_\beta Ln\left(\frac{-t}{\eta}\right)-\frac{\beta}{\eta}\sigma_\eta\right)\varepsilon \quad \text{[A4.27]}$$

So, we have:

$$Ln(R-1)+\frac{\sigma_R}{R-1}\varepsilon=\beta Ln\left(\frac{-t}{\eta}\right)+\left(\sigma_\beta Ln\left(\frac{-t}{\eta}\right)-\frac{\beta}{\eta}\sigma_\eta\right)\varepsilon \quad \text{[A4.28]}$$

The identification of the different terms in the last equation leads to extracting the mean and standard deviation of the random reliability as:

$$Ln(R-1) = \beta Ln\left(\frac{-t}{\eta}\right) \quad \text{[A4.29]}$$

$$\frac{\sigma_R}{R-1} = \sigma_\beta Ln\left(\frac{-t}{\eta}\right) - \frac{\beta}{\eta}\sigma_\eta \quad \text{[A4.30]}$$

And then:

$$R(t) = exp\left(\frac{-t}{\eta}\right)^\beta \quad \text{[A4.31]}$$

$$\sigma_R(t) = (R(t)-1)\left(\sigma_\beta Ln\left(\frac{-t}{\eta}\right) - \frac{\beta}{\eta}\sigma_\eta\right) \quad \text{[A4.32]}$$

References

Abd Wahab, N.H., Hasikin, K., Lai, K.W., Xia, K., Taing, A.Y., Zhang, R. (2025). Predicting medical device life expectancy and estimating remaining useful life using a data-driven multimodal framework. *IEEE Access*, 13, 117300–117327. doi: 10.1109/ACCESS.2025.3581548.

Ahsan, S., Lemma, T.A., Gebremariam, M.A. (2019). Reliability analysis of gas turbine engine by means of bathtub-shaped failure rate distribution. *Process Safety Progress*, 39(3). doi: 10.1002/prs.12115.

Al-Najjar, B. and Alsyouf, I. (2004). Enhancing a company's profitability and competitiveness using integrated vibration-based maintenance: A case study. *European Journal of Operational Research*, 157(3), 643–657. doi: 10.1016/S0377-2217(03)00258-3.

Alfares, H. (1999). A simulation model for determining inspection frequency. *Computers & Industrial Engineering*, 36(3), 685–696. doi: 10.1016/S0360-8352(99)00159-X.

Alkhatib, S.A., Katmah, R., Kosaji, D., Afzal, S.U.B., Tariq, M.H., Simsekler, M.C.E., Ellahham, S. (2025). AI-driven decision support framework for preventing medical equipment failure and enhancing patient safety: A new perspective. *J Multidiscip Healthcare*, 18, 6299–6313. doi: 10.2147/JMDH.S528612.

Alnazier, O.H., Osman, M.O., Khider, M.O. (2013). Evaluation of mechanical and electrical faults in dialysis machines. *Journal of Physical Therapy and Health Promotion*, 1(1), 1–7. doi: 10.18005/PTHP0101001.

Alvarenga, M.A.B., Frutuoso e Melo, P.F., Fonseca, R.A. (2014). A critical review of methods and models for evaluating organizational factors in human reliability analysis. *Progress in Nuclear Energy*, 75, 25–41. doi: 10.1016/j.pnucene.2014.04.004.

Annis, D.H. (2006). Kendall's advanced theory of statistics, Vol. 1: Distribution theory, Kendall's advanced theory of statistics, Vol. 2A: Classical inference and the linear model. *Journal of the American Statistical Association*, 101(476), 1721. doi: 10.1198/jasa.2006.s140.

Arunraj, N.S. and Maiti, J. (2007). Risk-based maintenance–Techniques and applications. *Journal of Hazardous Materials*, 142(3), 653–661. doi: 10.1016/j.jhazmat.2006.06.069.

Arya, V., Deshmukh, S.G., Bhatnagar, N. (2019). Performance analysis of high technology collaborative networks: A case of medical device manufacturing. *Computers & Industrial Engineering*, 137, 106065. doi: 10.1016/j.cie.2019.106065.

Aupied, J. (1993). *Retour d'expérience appliqué à la sûreté de fonctionnement des matériels en exploitation*. Éditions Eyrolles, EDF, Paris.

Aven, T., Sklet, S., Vinnem, J.E. (2006). Barrier and operational risk analysis of hydrocarbon releases (BORA-Release): Part I. Method description. *Journal of Hazardous Materials*, 137(2), 681–691. doi: 10.1016/j.jhazmat.2006.03.049.

Azar, A.T. (2009). The influence of maintenance quality of hemodialysis machines on hemodialysis efficiency. *Saudi J Kidney Dis Transpl*, 20(1), 49–56.

Bahreini, A.F. (2024). Ranking the states most at risk of healthcare data breaches: An application of integrated multicriteria framework for prioritization in risk management. *Organizational Cybersecurity Journal: Practice, Process & People*, 4(2), 53–84. doi: 10.1108/OCJ-01-2023-0001.

Bahreini, R., Doshmangir, L., Imani, A. (2018). Factors affecting medical equipment maintenance management: A systematic review. *Journal of Clinical and Diagnostic Research*, 12(4), 1–7. doi: 10.7860/JCDR/2018/31646.11375.

Balakrishnan, N. (ed.) (2015). Fault tree analysis for medical applications. In *Dependability in Medicine and Neurology*. Springer, Cham. doi: 10.1007/978-3-319-14968-4_3.

Bernal, J. and Mazo, C. (2022). Transparency of artificial intelligence in healthcare: Insights from professionals in computing and healthcare worldwide. *Applied Sciences*, 12(20), 10228. doi: 10.3390/app122010228.

Bertsche, B. (ed.) (2008). Maintenance and reliability. In *Reliability in Automotive and Mechanical Engineering*. Springer, Berlin, Heidelberg. doi: 10.1007/978-3-540-34282-3_10.

Bouchoucha, F., Ichchou, M.N., Haddar, M. (2017). Stochastic wave finite element method in uncertain elastic media through the second order perturbation. *J. Appl. Mech. Tech. Phy.*, 58, 362–370. doi: 10.1134/S0021894417020225.

Brissaud, F., Lanternier, B., Charpentier, D., Lyonnet, P. (2007). Modélisation des taux de défaillance en mécanique – Combinaison d'une loi de Weibull et d'un modèle de Cox pour la modélisation des taux de défaillance en fonction du temps et des facteurs d'influence. In *Performance et Nouvelles Technologies en Maintenance, PENTOM 2007*, Mons [Online]. Available at: https://hal.science/hal-00196765.

Brissaud, F., Charpentier, D., Fouladirad, M., Barros, A., Bérenguer, C. (2010). Failure rate evaluation with influencing factors. *Journal of Loss Prevention in the Process Industries*, 23(2), 187–193. doi: 10.1016/j.jlp.2009.07.013.

Cao, H., Yu, J., Duan, F. (2025). Condition-based maintenance in complex degradation systems: A review of modeling evolution, multi-component systems, and maintenance strategies. *Machines*, 13(8), 714. doi: 10.3390/machines13080714.

Chatzimichail, T. and Hatjimihail, A.T. (2023). A Bayesian inference based computational tool for parametric and nonparametric medical diagnosis. *Diagnostics*, 13(19), 3135. doi: 10.3390/diagnostics13193135.

Chesley, J.C. and Naval Surface Warfare Center Logistics Engineering Technology Branch (1998). *Handbook of Reliability Prediction Procedures for Mechanical Equipment*. NSWC Carderock Division, West Bethesda, MD [Online]. Available at: https://books.google.tn/books?id=iyWoPgAACAAJ.

Chowdhuri, I.R., Kar, A., Pal, A.K. (2023). A new approach for effective reliability management of biomedical equipment. *Applied Sciences*, 1(5), 281–293. doi: 10.59324/ejtas.2023.1(5).19.

Cohen, C.A. and Whitten, B. (1982). Modified maximum likelihood and modified moment estimators for the three-parameter Weibull distribution. *Communications in Statistics – Theory and Methods*, 11(23), 2631–2656. doi: 10.1080/03610928208828412.

Cui, T.J. and Li, S.-S. (2019). Study on the relationship between system reliability and influencing factors under big data and multi-factors. *Cluster Computing*, 22(4), 10275–10297. doi: 10.1007/s10586-017-1278-5.

David, Y. and Jahnke, E.G. (2004). Planning hospital medical technology management. *IEEE Engineering in Medicine and Biology Magazine*, 23(3), 73–79. doi: 10.1109/MEMB.2004.1317985.

Davis, S.E., Greevy, R.A., Jr., Lasko, T.A., Walsh, C.G., Matheny, M.E. (2020). Detection of calibration drift in clinical prediction models to inform model updating. *Journal of Biomedical Informatics*, 112, 103611. doi: 10.1016/j.jbi.2020.103611.

Deloux, E., Castanier, B., Bérenguer, C. (2008). Maintenance policy for a deteriorating system evolving in a stressful environment. *Proceedings of the Institution of Mechanical Engineers, Part O: Journal of Risk and Reliability*, 222(4), 613-622. doi: 10.1243/1748006XJRR148.

Devooght, J. (2002). Dynamic reliability. In *Advances in Nuclear Science and Technology*, Lewins, J. and Becker, M. (eds). Springer, Boston, MA, 215–278. doi: 10.1007/0-306-47812-9_7.

Dhillon, B.S. (2000). *Medical Device Reliability and Associated Areas*. CRC Press, Boca Raton, FL. doi: 10.1201/9781420042238.

Dhillon, B.S. (2006). *Maintainability, Maintenance and Reliability for Engineers*. CRC Press, Boca Raton, FL. doi: 10.1201/9781420006780.

Dhillon, B.S. (2011). Medical equipment reliability: A review, analysis methods and improvement strategies. *International Journal of Reliability, Quality and Safety Engineering*, 18(4), 391–403. doi: 10.1142/S0218539311004317.

European Commission (2017). Regulation (EU) 2017/745 on medical devices. *Official Journal of the European Union* [Online]. Available at: http://data.europa.eu/eli/reg/2017/745/oj.

Fenina, S. and Bouchoucha, F. (2023). Weibull distribution analysis of the hemodialysis machine reliability: A case study. In *Proceedings of the 2nd International Conference on Innovative Materials, Manufacturing, and Advanced Technologies. IMMAT 2022.* Springer, Cham. doi: 10.1007/978-3-031-42659-9_19.

Fenina, S., Jendoubi, S., Bouchoucha, F. (2023). Failure rate estimation by Weibull distribution in a stochastic environment: Application to the hemodialysis machine. *Reliability: Theory and Application*, 18, 450–463. doi: 10.24412/1932-2321-2023-374-450-463.

Fenina, S., Jendoubi, S., Bouchoucha, F. (2024a). Effect of the stochastic weibull parameters on the reliability of the dialysis machine. *Engineering and Technology Journal*, 42(4), 407–420. doi: 10.30684/etj.2024.145834.1665.

Fenina, S., Jendoubi, S., Bouchoucha, F. (2024b). Reliability of the hemodialysis machine through Weibull distribution in the presence of uncertainties. *Industrial and Systems Engineering Journal*, 7(1). doi: 10.21494/ISTE.OP.2024.1108.

Ferreira, L.A. and Silva, J.L. (2017). Parameter estimation for Weibull distribution with right censored data using EM algorithm. *Eksploatacja i Niezawodnosc – Maintenance and Reliability*, 19(2), 310–315. doi: 10.17531/ein.2017.2.20.

Fidanoglu, M., Ungor, U., Ozkol, I., Komurgoz, G. (2017). Application of Weibull distribution method for aircraft component life estimation in civil aviation sector. *Journal of Traffic and Logistics Engineering*, 5(1), 40–44. doi: 10.18178/jtle.5.1.40-44.

Frolov, M., Tanchenko, S., Ohluzdina, L. (2024). Parameter estimation of the Weibull distribution in modeling the reliability of technical objects. *Journal of Engineering Sciences (Ukraine)*, 11(1), 1–10. doi: 10.21272/jes.2024.11(1).a1.

Gatouillat, A., Badr, Y., Massot, B., Sejdić, E. (2018). Internet of medical things: A review of recent contributions dealing with cyber-physical systems in medicine. *IEEE Internet of Things Journal*, 5(5), 3810–3822. doi: 10.1109/JIOT.2018.2849014.

Gölbaşı, O. and Demirel, N. (2014). Stochastic models in preventive maintenance policies. *Advanced Materials Research*, 1016, 802. doi: 10.4028/www.scientific.net/AMR.1016.802.

Gorjian, N., Ma, L., Mittinty, M., Yarlagadda, P., Sun, Y. (2010). A review on degradation models in reliability analysis. In *Engineering Asset Lifecycle Management*, Kiritsis, D., Emmanouilidis, C., Koronios, A., Mathew, J. (eds). Springer, London. doi: 10.1007/978-0-85729-320-6_42.

Gouno, E. and Guérineau, L. (2015). Failure rate estimation in a dynamic environment. *Economic Quality Control*, 30(1), 1–8. doi: 10.1515/eqc-2015-6001.

Habchi, G. and Barthod, C. (2016). An overall methodology for reliability prediction of mechatronic systems design with industrial application. *Reliability Engineering & System Safety*, 155, 236–254. doi: 10.1016/j.ress.2016.06.013.

Hadroug, N., Hafaifa, A., Kouzou, A., Chaibet, A. (2016). Improvement of gas turbine availability using reliability modeling based on fuzzy system. In *International Congress on Technical Diagnostics*. Springer, Gliwice. doi: 10.1007/978-3-319-62042-8_2.

Hajji, A. (2019). Evaluation des risques au niveau des établissements d'hémodialyse au Maroc : application d'une démarche AMDEC. Master's Thesis, Ecole Nationale de Commerce et de Gestion de Casablanca. doi: 10.13140/RG.2.2.13828.14720.

Hsu, F.-Y., Hwang, R.-H., Tsai, M.-H., Wang, J.-T. (2024). Predicting peritoneal dialysis failure within the next three months based on deep learning and important features analysis. *Information*, 15(12), 776. doi: 10.3390/info15120776.

Hu, J. and Chen, P. (2020). Predictive maintenance of systems subject to hard failure based on proportional hazards model. *Reliability Engineering & System Safety*, 196, 106707. doi: 10.1016/j.ress.2019.106707.

Hunte, J.L., Neil, M., Fenton, N. E. (2024). A hybrid Bayesian network for medical device risk assessment and management. *Reliability Engineering & System Safety*, 241, 109630. doi: 10.1016/j.ress.2023.109630.

Hyman, W.A. and Johnson, E. (2008). Fault tree analysis of clinical alarms. *Journal of Clinical Engineering*, 33(2), 85–94. doi: 10.1097/01.JCE.0000305872.86942.66.

Ianculescu, D. and Anghel, C. G. (2025). Innovative explicit relations for Weibull distribution parameters based on K-moments. *Mathematics*, 13(21), 3473. doi: 10.3390/math13213473.

Ichchou, M.N., Bouchoucha, F., Ben Souf, M.A., Dessombz, O., Haddar, M. (2011). Stochastic wave finite element for random periodic media through first-order perturbation. *Comput. Methods Appl. Mech. Engrg.*, 200, 2805–2813. doi: 10.1016/j.cma.2011.05.004.

Jamshidi, A., Rahimi, S.A., Ait-Kadi, D., Bartolome, A.R. (2014). Medical devices inspection and maintenance; a literature review. In *IISE Annual Conference. Proceedings*. Institute of Industrial and Systems Engineers (IISE) [Online]. Available at: https://www.academia.edu/download/104336744/I1418.pdf.

Jaros, R., Byrtus, R., Dohnal, J., Danys, L., Baros, J., Koziorek, J., Zmij, P., Martinek, R. (2023). Advanced signal processing methods for condition monitoring. *Arch Computat Methods Eng*, 30, 1553–1577. doi: 10.1007/s11831-022-09834-4.

Jiménez, J. J. M., Vingerhoeds, R. A., Grabot, B., Schwartz, S. (2023). An ontology model for maintenance strategy selection and assessment. *Journal of Intelligent Manufacturing*, 34, 1369–1387. doi: 10.1007/s10845-021-01855-3.

Kappenman, R.F. (1985). Estimation for the three-parameter Weibull, lognormal, and gamma distributions. *Computational Statistics & Data Analysis*, 3, 11–23. doi: 10.1016/0167-9473(85)90054-4.

Khalaf, A. and Hamam, Y. (2013). The effect of maintenance on the survival of medical equipment. *Journal of Engineering, Design and Technology*, 11(2), 142–157. doi: 10.1108/JEDT-06-2011-0033.

Khalaf, A., Djouani, K., Hamam, Y., Alayli, Y. (2010). Evidence-based mathematical maintenance model for medical equipment. In *Proceedings of the International Conference on Electronic Devices, Systems and Applications (ICEDSA 2010)*, Kuala Lumpur. IEEE, Piscataway, NJ. doi: 10.1109/ICEDSA.2010.5503071.

Khan, S., Hasan, O., Mashkoor, A. (2018). Formal verification and safety assessment of a hemodialysis machine. In *SOFSEM 2018: Theory and Practice of Computer Science*. Edizioni della Normale, Cham. doi: 10.1007/978-3-319-73117-9_17.

Kim, J.S. and Yum, B.-J. (2008). Selection between Weibull and lognormal distributions: A comparative simulation study. *Computational Statistics & Data Analysis*, 53(2), 477–485. doi: 10.1016/j.csda.2008.08.012.

Kotanko, P., Zhang, H., Wang, Y. (2023). Artificial intelligence and machine learning in dialysis: Ready for prime time? *Clinical Journal of the American Society of Nephrology: CJASN*, 18(6), 803–805. doi: 10.2215/CJN.0000000000000089.

Kremer, T., Murray, N., Buckley, J., Rowan, N.J. (2023). Use of real-time immersive digital training and educational technologies to improve patient safety during the processing of reusable medical devices: Quo Vadis? *Science of The Total Environment*, 900, 165673. doi: 10.1016/j.scitotenv.2023.165673.

Krit, M., Gaudoin, O., Remy, E. (2021). Goodness-of-fit tests for the Weibull and extreme value distributions: A review and comparative study. *Communications in Statistics – Simulation and Computation*, 50(7), 1888–1911. doi: 10.1080/03610918.2019.1594292.

Langseth, H. and Portinale, L. (2007). Bayesian networks in reliability. *Reliability Engineering & System Safety*, 92(1), 92–108. doi: 10.1016/j.ress.2005.11.037.

Le Son, K., Fouladirad, M., Barros, A., Levrat, E., Iung, B. (2013). Remaining useful life estimation based on stochastic deterioration models: A comparative study. *Reliability Engineering & System Safety*, 112, 165–175. doi: 10.1016/j.ress.2012.11.022.

Lee, J., Bagheri, B., Kao, H.A. (2015). A cyber-physical systems architecture for industry 4.0-based manufacturing systems. *Manufacturing Letters*, 3, 18–23. doi: 10.1016/j.mfglet.2014.12.001.

Leming, M. J., Bron, E. E., Bruffaerts, R., Ou, Y., Iglesias, J.E., Gollub, R.L., Im, H. (2023). Challenges of implementing computer-aided diagnostic models for neuroimages in a clinical setting. *NPJ Digital Medicine*, 6(1), 129. doi: 10.1038/s41746-023-00868-x.

Li, M., Liu, Z., Li, X., Liu, Y. (2019). Dynamic risk assessment in healthcare based on Bayesian approach. *Reliability Engineering & System Safety*, 189, 327–334. doi: 10.1016/j.ress.2019.04.040.

Liao, H.Y., Cade, W., Behdad, S. (2021). Markov chain optimization of repair and replacement decisions of medical equipment. *Resources, Conservation and Recycling*, 171, 105609. doi: 10.1115/1.4051886.

Lihou, D.A. and Spence, G.D. (1988). Proper use of data with the Weibull distribution. *Journal of Loss Prevention in the Process Industries*, 1(2), 110–113. doi: 10.1016/0950-4230(88)80021-4.

Lin, C.H., Ho, T.F., Chen, H.F., Chang, H.Y., Chien, J.H. (2022). Applying healthcare failure mode and effect analysis and the development of a real-time mobile application for modified early warning score notification to improve patient safety during hemodialysis. *Journal of Patient Safety*, 18(5), 475–485. doi: 10.1097/PTS.0000000000000977.

Lin, Z., Kang, J., Wei, Y., Zou, B. (2024). Maintenance management strategies for medical equipment in healthcare institutions: A review. *BME Horiz*, 2, 2024135. doi: 10.70401/bmeh.2024.135.

Liu, Y., Berg, R., Chen, X., Abeyratne, A., Wang, X., Haddad, T. (2015). Bayesian reliability prediction of a medical device system. In *Proceedings of the First International Conference on Reliability Systems Engineering (ICRSE 2015)*, Beijing. IEEE, Piscataway, NJ. doi: 10.1109/ICRSE.2015.7366449.

Lodi, C.A., Vasta, A., Hegbrant, M.A., Bosch, J.P., Paolini, F., Garzotto, F., Ronco, C. (2010). Multidisciplinary evaluation for severity of hazards applied to hemodialysis devices: An original risk analysis method. *Clin J Am Soc Nephrol.*, 5(11), 2004–2017. doi: 10.2215/CJN.01740210.

Luko, S.N. (1999). A review of the Weibull distribution and selected engineering applications. SAE Technical Paper Series, 1999-01-2859. doi: 10.4271/1999-01-2859.

Lyonnet, P. (1991). *Tools of Total Quality: An Introduction to Statistical Process Control*. Chapman & Hall, London. doi: 10.1007/978-94-011-3140-7.

Madhukar, M. (2005). The basics of hemodialysis equipment. *Hemodialysis International*, 9(1), 30–36. doi: 10.1111/j.1492-7535.2005.01115.x.

Mazah, M.M. (2019). Efficiency of the reliability function and scheduling method in determining the optimal preventive maintenance time for digital X-ray devices. *International Journal of Statistics and Applications*, 9, 53–58 [Online]. Available at: https://api.semanticscholar.org/CorpusID:182030028.

McCauley, N. and Ala, M. (1992). The use of expert systems in the healthcare industry. *Information & Management*, 22(4), 227–235. doi: 10.1016/0378-7206(92)90025-B.

McLachlan, S., Dube, K., Hitman, G.A., Fenton, N.E., Kyrimi, E. (2020). Bayesian networks in healthcare: Distribution by medical condition. *Artificial Intelligence in Medicine*, 107, 101912. doi: 10.1016/j.artmed.2020.101912.

Meijer, C., Uh, H.W., El Bouhaddani, S. (2023). Digital twins in healthcare: Methodological challenges and opportunities. *Journal of Personalized Medicine*, 13(10), 1522. doi: 10.3390/jpm13101522.

Mengual, E.S., López, P.P., García, S.G., Lozano, R.G., Feijoo, G., Moreira, M.T., Gabarrell, X., Rieradevall, J. (2014). Eco-designing the use phase of products in sustainable manufacturing. *Journal of Industrial Ecology*, 18(4), 545–575. doi: 10.1111/jiec.12161.

Mijailovic, V. (2003). Probabilistic method for planning of maintenance activities of substation components. *Electric Power Systems Research*, 64(1), 53–58. doi: 10.1016/S0378-7796(02)00148-7.

Mkalaf, K.A. (2016). A study of current maintenance strategies and the reliability of critical medical equipment in hospitals in relation to patient outcomes. Thesis, University of Wollongong [Online]. Available at: https://hdl.handle.net/10779/uow.27660966.v1.

Mutia, D.M., Mukhongo, L., Chemweno, P. (2018). Review in maintenance strategies for haemodialysis machine in healthcare facilities. *Industrial Engineering*, 2(1), 34–41. doi: 10.11648/j.ie.20180201.15.

Namuduri, S., Narayanan, B.N., Davuluru, V.S.P., Burton, L., Bhansali, S. (2020). Deep learning methods for sensor based predictive maintenance and future perspectives for electrochemical sensors. *Journal of The Electrochemical Society*, 167(3), 037552. doi: 10.1149/1945-7111/ab67a8.

O'Connor, D.T. and Kleyner, A. (2012). *Practical Reliability Engineering*, 5th edition. Wiley, Hoboken, NJ. doi: 10.1002/9781119961260.

Ouakki, W. (2011). Prise en compte de la fiabilité en conception. Master's Thesis, Université Laval, Quebec [Online]. Available at: http://hdl.handle.net/20.500.11794/23131.

Pasha, G.R., Khan, S., Pasha, A.H. (2006). Empirical analysis of the Weibull distribution for failure data. *Journal of Statistics*, 13(1), 33–45 [Online]. Available at: https://jstatgcu.pk/index.php/jstat/article/view/149.

Pestana, M., Pereira, R., Moro, S. (2020). Improving health care management in hospitals through a productivity dashboard. *J. Med. Syst.*, 44, 87. doi: 10.1007/s10916-020-01546-1.

Rausand, M., Barros, A., Hoyland, A. (2004). *System Reliability Theory: Models, Statistical Methods, and Applications*, 2nd edition. Wiley, Hoboken, NJ [Online]. Available at: https://books.google.tn/books?id=gkUWz9AA-QEC.

Rausand, M., Barros, A., Hoyland, A. (2020). *System Reliability Theory: Models, Statistical Methods, and Applications*, 3rd edition. Wiley, Hoboken, NJ. doi: 10.1002/9781119373940.

Rudsari, H.K., Tseng, B., Zhu, H., Song, L., Gu, C., Roy, A., Irajizad, E., Butner, J., Long, J., Do, K.A. (2025). Digital twins in healthcare: A comprehensive review and future directions. *Frontiers in Digital Health*, 7, 1633539. doi: 10.3389/fdgth.2025.1633539.

Ruijters, E. and Stoelinga, M. (2015). Fault tree analysis: A survey of the state of the art in modeling, analysis and tools. *Computer Science Review*, 15–16, 29–62. doi: 10.1016/j.cosrev.2015.03.001.

Ronco, C. and La Greca, G. (2002). The role of technology in hemodialysis. *Contrib Nephrol*, 137, 1–12. doi: 10.1159/000060259.

Sabah, S., Moussa, M., Shamayleh, A. (2022). Predictive maintenance application in healthcare. In *Annual Reliability and Maintainability Symposium (RAMS)*, Tucson, AZ. doi: 10.1109/RAMS51457.2022.9893973.

Sabzaliyev, A. (2024). Knowledge representation in expert systems: Structure, classification, and applications. *Luminis Applied Science and Engineering*, 1(2), 1–15. doi: 10.69760/lumin.202400005.

Saleh, N., Sharawi, A.A., Elwahed, M.A., Petti, A., Puppato, D., Balestra, G. (2014). Preventive maintenance prioritization index of medical equipment using quality function deployment. *IEEE J. Biomed Health Inform.*, 19(3), 1029–1035. doi: 10.1109/JBHI.2014.2337895.

Saraswat, S. and Yadava, G. (2008). An overview on reliability, availability, maintainability and supportability (RAMS) engineering. *International Journal of Quality & Reliability Management*, 25(3), 330–344. doi: 10.1108/02656710810854313.

Schober, P., Boer, C., Schwarte, L.A. (2018). Correlation coefficients: Appropriate use and interpretation. *Anesthesia & Analgesia*, 126(5), 1763–1768. doi: 10.1213/ANE.0000000000002864.

Shahriari, M., Shahrasbi, H., Zaretalab, A. (2024). Reliability analysis of lifetime systems based on Weibull distribution. *International Journal of Nonlinear Analysis and Applications*, 15(1), 321–329. doi: 10.22075/ijnaa.2024.31309.4609.

Shahzad, M.N., Ullah, E., Hussanan, A. (2019). Beta exponentiated modified Weibull distribution: Properties and application. *Symmetry*, 11(6), 781. doi: 10.3390/sym11060781.

Shamayleh, A., Awad, M., Abdulla, A.O. (2020). Criticality-based reliability-centered maintenance for healthcare. *Journal of Quality in Maintenance Engineering*, 26(2), 311–334. doi: 10.1108/JQME-10-2018-0084.

Sielaffa, L. and Lucke, D. (2021). An approach for an integrated maintenance strategy selection considering the context of the value-adding network. *Procedia CIRP*, 104, 815–820. doi: 10.1016/j.procir.2021.11.137.

Silva, M.S., Nunes, M.A.A., Rosa, S.S.R.F., Sousa, M.M., Santos, R.P., Farias, I.P., Lima, F.J.M., Correia, S.D. (2025). Reliability analysis of hospital infusion pumps: A case study. *BioMedical Engineering Online*, 24(1), 55. doi: 10.1186/s12938-025-01389-2.

Stawowy, M., Rosiński, A., Paś, J., Klimczak, T. (2021). Method of estimating uncertainty as a way to evaluate continuity quality of power supply in hospital devices. *Energies*, 14(2), 486. doi: 10.3390/en14020486.

Stone, G.C. and Van Heeswijk, R.G. (2007). Parameter estimation for the Weibull distribution. *IEEE Transactions on Electrical Insulation*, 4, 253–261. doi: 10.1109/TEI.1977.297976.

Sundaramurthy, A. and Vaithiyalingam, C. (2025). Advancements in calibration techniques for ensuring accuracy and reliability of medical devices: A comprehensive report. In *Hybrid and Advanced Technologies*, Christydass, S.P.J., Nurhayati, N., Kannadhasan, S. (eds). CRC Press, Boca Raton, FL. doi: 10.1201/9781003559139-60.

Taghipour, S., Banjevic, D., Jardine, A.K.S. (2011). Prioritization of medical equipment for maintenance decisions. *Journal of the Operational Research Society*, 62(9), 1666–1687. doi: 10.1057/jors.2010.106.

Tambe, S.N., Potharaju, S., Amiripalli, S.S., Tirandasu, R.K., Jaidhan, B.J. (2025). Interdisciplinary research for predictive maintenance of MRI machines using machine learning. *Journal of Current Science and Technology*, 15(1), 78. doi: 10.59796/jcst.V15N1.2025.78.

Taştan M. (2025). Machine learning-based calibration and performance evaluation of low-cost internet of things air quality sensors. *Sensors*, 25(10), 3183. doi: 10.3390/s25103183.

Tebbi, O., Guérin, F., Dumon, B. (2005). Estimation de la fiabilité par les essais accélérés. *Mécanique & Industries*, 6(2), 155–167. doi: 10.1051/meca:2005016.

Tonelli, D., Beltempo, A., Cappello, C., Bursi, O.S., Zonta, D. (2023). Reliability analysis of complex structures based on Bayesian inference. *Structural Health Monitoring*, 22(5), 3481–3497. doi: 10.1177/14759217231152798.

Tsai, C.-L., Fehintola, A., Crooijmans, G., Vollenbroek, J., Brett Duane, B., Gerritsen, K. (2025). Sustainable and accessible hemodialysis: Life cycle assessment on central acid delivery system. *BMC Nephrol*, 26, 619. doi: 10.1186/s12882-025-04499-0.

Tsarouhas, P. (2012). Reliability analysis into hospital dialysis system: A case study. *Quality and Reliability Engineering International*, 29(8), 1235–1243. doi: 10.1002/qre.1468.

Tsarouhas, P. (2019). Statistical analysis of failure data for estimating reliability, availability and maintainability of an automated croissant production line. *Journal of Quality in Maintenance Engineering*, 25(3), 452–475. doi: 10.1108/JQME-04-2018-0029.

Tully, J., Selzer, J., Phillips, J.P., O'Connor, P., Dameff, C. (2020). Healthcare challenges in the era of cybersecurity. *Health Security*, 18(3), 228–231. doi: 10.1089/hs.2019.0123.

Valdez-Flores, C. and Feldman, R.M. (1989). A survey of preventive maintenance models for stochastically deteriorating single-unit systems. *Naval Research Logistics*, 36, 419–446. doi: 10.1002/1520-6750(198908)36:4<419::AID-NAV3220360407>3.0.CO;2-5.

Vesely, W.E., Goldberg, F.F., Roberts, N.H., Haasl, D.F. (1981). *Fault Tree Handbook (NUREG-0492)*. U.S. Nuclear Regulatory Commission, Washington, D.C. [Online]. Available at: https://www.nrc.gov/docs/ML1007/ML100780465.pdf.

Wang, H. and Pham, H. (2006). *Reliability and Optimal Maintenance*. Springer, London. doi: 10.1007/B138077.

Wang, B., Rui, T., Skinner, S., Ayers-Comegys, M., Gibson, J., Williams, S. (2024). Medical equipment aging: Part I–Impact on maintenance. *Journal of Clinical Engineering*, 49(2), 52–64. doi: 10.1097/JCE.0000000000000644.

Weber, P., Medina-Oliva, G., Simon, C., Iung, B. (2022). Overview on Bayesian networks applications for dependability, risk analysis and maintenance areas. *Engineering Applications of Artificial Intelligence*, 25(4), 671–682. doi: 10.1016/j.engappai.2010.06.002.

Williams, P.A. and Woodward, A.J. (2015). Cybersecurity vulnerabilities in medical devices: A complex environment and multifaceted problem. *Medical Devices: Evidence and Research*, 8, 305–316. doi: 10.2147/MDER.S50048.

World Health Organization (2025). *Development of Medical Device Policies*. World Health Organization, Geneva.

World Health Organization, Regional Committee for the Eastern Mediterranean (2006). The role of medical devices and equipment in contemporary health care systems and services. EM/RC53/Tech.Disc.2, Cairo [Online]. Available at: https://iris.who.int/handle/10665/122442.

Wu, S.-J., Gebraeel, N., Lawley, M.A., Yih, Y. (2007). A neural network integrated decision support system for condition-based optimal predictive maintenance policy. *IEEE Transactions on Systems, Man, and Cybernetics – Part A: Systems and Humans*, 37(2), 226–236. doi: 10.1109/TSMCA.2006.886368.

Wu, J., Chen, C., Ma, Y., Xiu, Z., Cheng, Z., Pan, Y., Song, S. (2025). Integrated fault tree and case analysis for equipment conventional fault IETM diagnosis. *Sensors*, 25(17), 5231. doi: 10.3390/s25175231.

Xie, S., Xue, F., Zhang, W., Zhu, J. (2023). Data-driven predictive maintenance policy based on dynamic probability distribution prediction of remaining useful life. *Machines*, 11(10), 923. doi: 10.3390/machines11100923.

Xu, Q., Xie, W., Liao, B., Hu, C., Qin, L., Yang, Z., Luo, A. (2023). Interpretability of clinical decision support systems based on artificial intelligence from technological and medical perspective: A systematic review. *Journal of Healthcare Engineering*, 1, 9919269. doi: 10.1155/2023/9919269.

Yu, S., Lee S.S., Hwang, H. (2024). The ethics of using artificial intelligence in medical research. *Kosin Med J*, 39(4), 229–237. doi: 10.7180/kmj.24.140.

Zamzam, A.H., Abdul Wahab, A.K., Azizan, M.M., Satapathy, S.C., Lai, K.W., Hasikin, K. (2021). A systematic review of medical equipment reliability assessment in improving the quality of healthcare services. *Frontiers in Public Health*, 9, 753951. doi: 10.3389/fpubh.2021.753951.

Zamzam, A.H., Hasikin, K., Wahab, A.K.A. (2023). Integrated failure analysis using machine learning predictive system for smart management of medical equipment maintenance. *Engineering Applications of Artificial Intelligence*, 125, 106715. doi: 10.1016/j.engappai.2023.106715.

Zhang, T. and Xie, M. (2007). Failure data analysis with extended Weibull distribution. *Communications in Statistics – Simulation and Computation*, 36(3), 579–592. doi: 10.1080/03610910701236081.

Zio, E. (2016). Some challenges and opportunities in reliability engineering. *IEEE Transactions on Reliability*, 65(4), 1769–1782. doi: 10.1109/TR.2016.2591504.

Zou K.H., Tuncali K., Silverman S.G. (2003). Correlation and simple linear regression. *Radiology*, 227(3), 617–622. doi: 10.1148/radiol.2273011499.

Index

K, L

M, O

P, R

S, U, W

Other titles from

ISTE

in

Mechanical Engineering and Solid Mechanics

2025

ABDELGHANI-IDRISSI Moulay, VERNIÈRES-HASSIMI Lamiae, ABDELGHANI-IDRISSI Soufiane
Thermal Convexity Analysis of a Solar Water Heating System

BAYLE Franck, DENIS Laurent, GIGLIATI Adrien
Monte Carlo Simulation in Dependability Analysis

EL HAMI Abdelkhalak
Methods and Applications of Artificial Intelligence: Dynamic Response, Learning, Random Forest, Linear Regression, Interoperability, Additive Manufacturing and Mechatronics
(Artificial Intelligence in Mechanics Set – Volume 2)

EL HAMI Abdelkhalak, RADI Bouchaïb
Multi-physics Optimization: Mechanics, Fluid Interaction Structure, Shaping, Stochastic Finite Elements, Random Vibrations, Fatigue
(Reliability of Multiphysical Systems Set – Volume 20)

LEROY Maurice
Rheology, Physical and Mechanical Behavior of Materials 4: Rigidity and Resistance of Composite Materials, Sizings of Laminate
Rheology, Physical and Mechanical Behavior of Materials 5: Fractures and Fatigue of Metal and Composites

2024

LE VAN Anh
Contact in Structural Mechanics: A Weighted Residual Approach

LEROY Maurice
Rheology, Physical and Mechanical Behavior of Materials 2: Working at High Strain Rates, Forming, Sintering, Joining and Welding
Rheology, Physical and Mechanical Behavior of Materials 3: Rigidity and Resistance of Materials, Sizings, Pieces and Structures

SAOUAB Abdelghani, CHAMPMARTIN Stéphane, KHALID NACIRI Jaafar
Some Complex Phenomena in Fluid and Solid Mechanics
(Mathematical and Mechanical Engineering Set – Volume 13)

2023

EL HAMI Abdelkhalak
Uncertainty and Artificial Intelligence: Additive Manufacturing, Vibratory Control, Agro-composite, Mechatronics
(Artificial Intelligence in Mechanics Set – Volume 1)

EL HAMI Abdelkhalak, EID Mohamed
Reliability-based Modeling of System Performance
(Reliability of Multiphysical Systems Set – Volume 19)

EL HAMI Abdelkhalak, DELAUX David, GRZESKOWIAK Henri
Applied Reliability for Industry 1: Predictive Reliability for the Automobile, Aeronautics, Defense, Medical, Marine and Space Industries
(Reliability of Multiphysical Systems Set – Volume 16)
Applied Reliability for Industry 2: Experimental Reliability for the Automobile, Aeronautics, Defense, Medical, Marine and Space Industries
(Reliability of Multiphysical Systems Set – Volume 17)
Applied Reliability for Industry 3: Operational Reliability for the Automobile, Aeronautics, Defense, Medical, Marine and Space Industries
(Reliability of Multiphysical Systems Set – Volume 18)

LEROY Maurice
Rheology, Physical and Mechanical Behavior of Materials 1: Physical Mechanisms of Deformation and Dynamic Behavior

2022

BAYLE Franck
Product Maturity 1: Theoretical Principles and Industrial Applications
(Reliability of Multiphysical Systems Set – Volume 12)
Product Maturity 2: Principles and Illustrations
(Reliability of Multiphysical Systems Set – Volume 13)

EL HAMI Abdelkhalak, DELAUX David, GRZESKOWIAK Henri
Reliability and Physics-of-Healthy in Mechatronics
(Reliability of Multiphysical Systems Set – Volume 15)

LANNOY André
Reliability of Nuclear Power Plants: Methods, Data and Applications
(Reliability of Multiphysical Systems Set – Volume 14)

LEDOUX Michel, EL HAMI Abdelkhalak
Heat Transfer 3: Convection, Fundamentals and Monophasic Flows
(Mathematical and Mechanical Engineering Set – Volume 11)
Heat Transfer 4: Convection, Two-Phase Flows and Special Problems
(Mathematical and Mechanical Engineering Set – Volume 12)

PLANCHETTE Guy
Cindynics, The Science of Danger: A Wake-up Call
(Reliability of Multiphysical Systems Set – Volume 11)

2021

CHALLAMEL Noël, KAPLUNOV Julius, TAKEWAKI Izuru
Modern Trends in Structural and Solid Mechanics 1: Static and Stability
Modern Trends in Structural and Solid Mechanics 2: Vibrations
Modern Trends in Structural and Solid Mechanics 3: Non-deterministic Mechanics

DAHOO Pierre Richard, POUGNET Philippe, EL HAMI Abdelkhalak
Applications and Metrology at Nanometer Scale 1: Smart Materials, Electromagnetic Waves and Uncertainties
(Reliability of Multiphysical Systems Set – Volume 9)
Applications and Metrology at Nanometer Scale 2: Measurement Systems, Quantum Engineering and RBDO Method
(Reliability of Multiphysical Systems Set – Volume 10)

LEDOUX Michel, EL HAMI Abdelkhalak
Heat Transfer 1: Conduction
(Mathematical and Mechanical Engineering Set – Volume 9)
Heat Transfer 2: Radiative Transfer
(Mathematical and Mechanical Engineering Set – Volume 10)

2020

SALENÇON Jean
Elastoplastic Modeling

2019

BAYLE Franck
Reliability of Maintained Systems Subjected to Wear Failure Mechanisms: Theory and Applications
(Reliability of Multiphysical Systems Set – Volume 8)

BEN KAHLA Rabeb, BARKAOUI Abdelwahed, MERZOUKI Tarek
Finite Element Method and Medical Imaging Techniques in Bone Biomechanics
(Mathematical and Mechanical Engineering Set – Volume 8)

IONESCU Ioan R., QUEYREAU Sylvain, PICU Catalin R., SALMAN Oguz Umut
Mechanics and Physics of Solids at Micro- and Nano-Scales

LE VAN Anh, BOUZIDI Rabah
Lagrangian Mechanics: An Advanced Analytical Approach

MICHELITSCH Thomas, PÉREZ RIASCOS Alejandro, COLLET Bernard, NOWAKOWSKI Andrzej, NICOLLEAU Franck
Fractional Dynamics on Networks and Lattices

SALENÇON Jean
Viscoelastic Modeling for Structural Analysis

VÉNIZÉLOS Georges, EL HAMI Abdelkhalak
Movement Equations 5: Dynamics of a Set of Solids
(Non-deformable Solid Mechanics Set – Volume 5)

2018

BOREL Michel, VÉNIZÉLOS Georges
Movement Equations 4: Equilibriums and Small Movements
(Non-deformable Solid Mechanics Set – Volume 4)

FROSSARD Etienne
Granular Geomaterials Dissipative Mechanics: Theory and Applications in Civil Engineering

RADI Bouchaib, EL HAMI Abdelkhalak
Advanced Numerical Methods with Matlab® 1: Function Approximation and System Resolution
(Mathematical and Mechanical Engineering Set – Volume 6)
Advanced Numerical Methods with Matlab® 2: Resolution of Nonlinear, Differential and Partial Differential Equations
(Mathematical and Mechanical Engineering Set – Volume 7)

SALENÇON Jean
Virtual Work Approach to Mechanical Modeling

2017

BOREL Michel, VÉNIZÉLOS Georges
Movement Equations 2: Mathematical and Methodological Supplements (Non-deformable Solid Mechanics Set – Volume 2)
Movement Equations 3: Dynamics and Fundamental Principle (Non-deformable Solid Mechanics Set – Volume 3)

BOUVET Christophe
Mechanics of Aeronautical Solids, Materials and Structures
Mechanics of Aeronautical Composite Materials

BRANCHERIE Delphine, FEISSEL Pierre, BOUVIER Salima, IBRAHIMBEGOVIC Adnan
From Microstructure Investigations to Multiscale Modeling: Bridging the Gap

CHEBEL-MORELLO Brigitte, NICOD Jean-Marc, VARNIER Christophe
From Prognostics and Health Systems Management to Predictive Maintenance 2: Knowledge, Traceability and Decision (Reliability of Multiphysical Systems Set – Volume 7)

EL HAMI Abdelkhalak, RADI Bouchaib
Dynamics of Large Structures and Inverse Problems (Mathematical and Mechanical Engineering Set – Volume 5)
Fluid-Structure Interactions and Uncertainties: Ansys and Fluent Tools (Reliability of Multiphysical Systems Set – Volume 6)

KHARMANDA Ghias, EL HAMI Abdelkhalak
Biomechanics: Optimization, Uncertainties and Reliability (Reliability of Multiphysical Systems Set – Volume 5)

LEDOUX Michel, EL HAMI Abdelkhalak
Compressible Flow Propulsion and Digital Approaches in Fluid Mechanics
(Mathematical and Mechanical Engineering Set – Volume 4)
Fluid Mechanics: Analytical Methods
(Mathematical and Mechanical Engineering Set – Volume 3)

MORI Yvon
Mechanical Vibrations: Applications to Equipment

2016

BOREL Michel, VÉNIZÉLOS Georges
Movement Equations 1: Location, Kinematics and Kinetics
(Non-deformable Solid Mechanics Set – Volume 1)

BOYARD Nicolas
Heat Transfer in Polymer Composite Materials

CARDON Alain, ITMI Mhamed
New Autonomous Systems
(Reliability of Multiphysical Systems Set – Volume 1)

DAHOO Pierre Richard, POUGNET Philippe, EL HAMI Abdelkhalak
Nanometer-scale Defect Detection Using Polarized Light
(Reliability of Multiphysical Systems Set – Volume 2)

DORMIEUX Luc, KONDO Djimédo
Micromechanics of Fracture and Damage
(Micromechanics Set – Volume 1)

EL HAMI Abdelkhalak, RADI Bouchaib
Stochastic Dynamics of Structures
(Mathematical and Mechanical Engineering Set – Volume 2)

GOURIVEAU Rafael, MEDJAHER Kamal, ZERHOUNI Noureddine
From Prognostics and Health Systems Management to Predictive Maintenance 1: Monitoring and Prognostics
(Reliability of Multiphysical Systems Set – Volume 4)

KHARMANDA Ghias, EL HAMI Abdelkhalak
Reliability in Biomechanics
(Reliability of Multiphysical Systems Set –Volume 3)

MOLIMARD Jérôme
Experimental Mechanics of Solids and Structures

RADI Bouchaib, EL HAMI Abdelkhalak
Material Forming Processes: Simulation, Drawing, Hydroforming and Additive Manufacturing
(Mathematical and Mechanical Engineering Set – Volume 1)

DE SAXCÉ Géry, VALLÉE Claude
Galilean Mechanics and Thermodynamics of Continua

2015

JEBAHI Mohamed, ANDRE Damien, TERREROS Inigo, IORDANOFF Ivan
Discrete Element Method to Model 3D Continuous Materials
Discrete Element Model and Simulation of Continuous Materials Behavior Set – Volume 1

JEBAHI Mohamed, DAU Frédéric, CHARLES Jean-Luc, IORDANOFF Ivan
Discrete-continuum Coupling Method to Simulate Highly Dynamic Multi-scale Problems: Simulation of Laser-induced Damage in Silica Glass
Discrete Element Model and Simulation of Continuous Materials Behavior Set – Volume 2

KARLIČIĆ Danilo, MURMU Tony, ADHIKARI Sondipon, MCCARTHY Michael
Non-local Structural Mechanics

SAB Karam, LEBÉE Arthur
Homogenization of Heterogeneous Thin and Thick Plates

2014

ATANACKOVIC M. Teodor, PILIPOVIC Stevan, STANKOVIC Bogoljub, ZORICA Dusan
Fractional Calculus with Applications in Mechanics: Vibrations and Diffusion Processes
Fractional Calculus with Applications in Mechanics: Wave Propagation, Impact and Variational Principles

CIBLAC Thierry, MOREL Jean-Claude
Sustainable Masonry: Stability and Behavior of Structures

ILANKO Sinniah, MONTERRUBIO Luis E., MOCHIDA Yusuke
The Rayleigh–Ritz Method for Structural Analysis

LALANNE Christian
Mechanical Vibration and Shock Analysis – 5-volume series – 3rd edition
Sinusoidal Vibration – Volume 1
Mechanical Shock – Volume 2
Random Vibration – Volume 3
Fatigue Damage – Volume 4
Specification Development – Volume 5

LEMAIRE Maurice
Mechanics and Uncertainty

2013

ADHIKARI Sondipon
Structural Dynamic Analysis with Generalized Damping Models: Analysis
Structural Dynamic Analysis with Generalized Damping Models: Identification

BAILLY Patrice
Materials and Structures under Shock and Impact

BASTIEN Jérôme, BERNARDIN Frédéric, LAMARQUE Claude-Henri
Non-smooth Deterministic or Stochastic Discrete Dynamical Systems: Applications to Models with Friction or Impact

EL HAMI Abdelkhalak, RADI Bouchaib
Uncertainty and Optimization in Structural Mechanics

KIRILLOV Oleg N., PELINOVSKY Dmitry E.
Nonlinear Physical Systems: Spectral Analysis, Stability and Bifurcations

LUONGO Angelo, ZULLI Daniele
Mathematical Models of Beams and Cables

SALENÇON Jean
Yield Design

2012

DAVIM J. Paulo
Mechanical Engineering Education

DUPEUX Michel, BRACCINI Muriel
Mechanics of Solid Interfaces

ELISHAKOFF Isaac *et al.*
Carbon Nanotubes and Nanosensors: Vibration, Buckling and Ballistic Impact

GRÉDIAC Michel, HILD François
Full-Field Measurements and Identification in Solid Mechanics

GROUS Ammar
Fracture Mechanics – 3-volume series
Analysis of Reliability and Quality Control – Volume 1
Applied Reliability – Volume 2
Applied Quality Control – Volume 3

RECHO Naman
Fracture Mechanics and Crack Growth

2011

KRYSINSKI Tomasz, MALBURET François
Mechanical Instability

LAMBERT Stéphane, NICOT François
Rockfall Engineering

SOUSTELLE Michel
An Introduction to Chemical Kinetics

2010

BREITKOPF Piotr, FILOMENO COELHO Rajan
Multidisciplinary Design Optimization in Computational Mechanics

DAVIM J. Paulo
Biotribology

PAULTRE Patrick
Dynamics of Structures

SOUSTELLE Michel
Handbook of Heterogenous Kinetics

2009

BERLIOZ Alain, TROMPETTE Philippe
Solid Mechanics using the Finite Element Method

LEMAIRE Maurice
Structural Reliability

2007

GIRARD Alain, ROY Nicolas
Structural Dynamics in Industry

GUINEBRETIÈRE René
X-ray Diffraction by Polycrystalline Materials

KRYSINSKI Tomasz, MALBURET François
Mechanical Vibrations

KUNDU Tribikram
Advanced Ultrasonic Methods for Material and Structure Inspection

SIH George C., NAÏT-ABDELAZIZ Moussa, VU-KHANH Toan
Particle and Continuum Aspects of Mesomechanics

Printed and bound by CPI Group (UK) Ltd, Croydon, CR0 4YY

13/09/2026

14961015-0002